AQUATIC TOXICOLOGY RESEARCH FOCUS

AQUATIC TOXICOLOGY RESEARCH FOCUS

ELIAS P. SVENSSON
EDITOR

Nova Science Publishers, Inc.
New York

For permission to use material from this book please contact us:
Telephone 631-231-7269; Fax 631-231-8175
Web Site: http://www.novapublishers.com

LIBRARY OF CONGRESS CATALOGING-IN-PUBLICATION DATA

Aquatic toxicology research focus / Elias P. Svensson (editor).
p. cm.
ISBN 978-1-60456-192-0 (hardcover)
1. Water--Pollution--Toxicology. I. Svensson, Elias P.
QH90.8.T68A675 2007
571.9'5--dc22

2007044378

Published by Nova Science Publishers, Inc. ✣ New York

CONTENTS

PREFACE

Aquatic toxicology is the study of the effects of manufactured chemicals and other anthropogenic and natural materials and activities on aquatic organisms at various levels of organization, from subcellular through individual organisms to communities and ecosystems. This book presents the latest research in this field from around the globe.

Chapter 1 - Reactive oxygen species (ROS) including superoxide anion ($O_2^{\bullet-}$), hydrogen peroxide (H_2O_2), hydroxyl radical ($^{\bullet}OH$) and their derivatives are continuously generated by different cell structures under aerobic conditions. They are catabolized by several specialized systems, which together with ROS generation create certain steady-state ROS concentrations. The situation when the balance between generation and degradation processes is disturbed in favor of the first, and the ROS steady-state concentrations are increased has been called "oxidative stress". Reactive oxygen species cause damage to almost all cellular constituents. In aquatic organisms oxidative stress may be developed naturally, for example, at fluctuations of environmental oxygen concentration, at temperature challenges and due to UV radiation, etc. In addition, many xenobiotics also increase production of ROS or compromise detoxification systems, or even both. For instance, some of them can either directly enter redox cycles or do it after biotransformation. Catabolism of certain xenobiotics by cytochrome system also can enhance ROS production as well as ions of metals with changeable valence. Inactivation of antioxidants or their exhaustion may also result in oxidative stress. Most studies to date show that toxicants can affect the redox balance via different mechanisms, which stimulate oxidative damage to cellular components, including proteins, lipids, nucleic acids etc. and exposure of cells to enhanced ROS levels leads to the dangerous intermediate production, mutagenesis or even death. The general description of ROS generation and development of oxidative stress in aquatic animals is analyzed with the focus on transition metal ions.

Chapter 2 - Recent studies have documented the presence of a wide variety of pharmaceuticals in the environment. Active pharmaceutical ingredients (APIs) can enter water systems via several routes, but municipal sewage effluent is considered to be one of the most important. Pharmaceuticals have been detected both in sewage influent and effluent, in surface water and groundwater and increasingly also in drinking water. Information on the long-term effects of the presence of APIs on aquatic life or human health is largely lacking. The aim of this chapter is to review the ecotoxicological work dealing with environmental contamination by pharmaceuticals, with particular regard to the fate and effects of APIs in urban water cycles, and to provide future perspectives and research needs. Pharmaceuticals are widespread contaminants, entering the environment from a myriad of scattered points. Patients, in the case of drugs for human use, or animals for veterinary drugs, are the main sources of contamination. The first step to be taken for a thorough environmental and human health risk assessment is to develop standardised methods for toxicological evaluation, as single compound and combined in mixtures of APIs that are representative for the real environment. *In vitro* and *in vivo* studies suggest that APIs, taken singularly or in combinations, and in concentrations close to those detected in the environment, may have ecotoxicological effects. Preliminary studies indicate that mixtures of APIs may also pose a risk for human health. Once thorough toxicological and ecotoxicological effects have been thoroughly investigated, APIs of particular environmental and toxicological concern can be identified. Sewage treatment processes are an important point in the control of contamination, but conventional treatment plants are generally not able to efficiently abate a substantial number of water-borne APIs. Several variables play a role, however, in the processes of sewage treatment, and could be specifically adjusted to improve the efficiency of API abatement, mitigating the potential environmental and human exposure.

Chapter 3 - In the last decade, the emphasis has shifted to the physical and microbiological quality of water sources as reports have shown that there are an increasing percentage of surface- and groundwater that are being contaminated with waterborne pathogens. Increased land use, rural development, informal settlements and inadequate infrastructure for water and sanitation have increased the risk of faecal contamination of these water sources. This is compounded by alterations in the hydrological cycle, associated with global climate change that increases the magnitude and frequency of run-off events. Waterborne diseases and epidemics result from the ingestion of faecally polluted water containing enteric bacterial pathogens, human-specific enteric viruses and protozoan parasites. Enteric bacteria have traditionally been used as indicators of faecal contamination of source, drinking and recreational water. However, research has demonstrated that these bacteria may not be appropriate indicators of pathogenic viruses and the need for reliable index or indicator organisms of these pathogens have been widely recognized. Studies conducted over the past couple of years have investigated, evaluated and compared bacteriophage and chemical assays as indicators of waterborne pathogens. In one study, the potential of F-RNA phage genotyping and chemical analysis of faecal sterols were evaluated to indicate if faecal pollution of water sources were of human or animal origin. When applied to environmental samples, F-RNA phage genotyping provided useful information in certain instances and confirmed the presence of either human or animal faecal pollution. Chemical assays yielded more conclusive results when F-RNA phage genotyping results obtained from river water samples were ambiguous. In another study F-RNA phage genotyping was used to determine the origin of faecal pollution in traditional and CDC water containers in rural settlements. Results showed no difference between the different containers; both human and animal phage subtypes were detected. Although the detection of each specific pathogen seems to be the ideal, this is not always possible due to expensive analysis and the need of well-equipped laboratories and skilled personnel. It can be concluded that bacterial and bacteriophage indicators are still the best alternative and most cost-effective tools currently available and will continue to play a major role in the determination of the microbial quality of water.

Chapter 4 - This paper reviewed impacts of currently used pesticides to aquatic organisms including fish, amphibian, aquatic invertebrates and algae in terms of acute, longer-term, joint effects, and side-effect on ecological levels. Pesticides showed diversity in their toxicity to aquatic species. Among them organophosphates and pyrethroids were more toxic to aquatic animals than the other pesticides, They were estimated to be moderately to highly toxic in most cases. Due to ubiquitous inter-specific diversity in sensitive, dubious relationship between acute and longer-term toxicity, and various direction of joint effect, a correct and complete risk assessment was hard to achieve by limited data. Sublethal concentration of pesticides might alter physical or biochemical parameters, impair self-protecting system, or change behavior of exposed aquatic species. Pesticides also caused alteration in species configuration and energy transfer in aquatic systems. Researches cited in this paper indicated that some of the pesticides could pose high risk to fish or invertebrates, even when they were applied at authorized rate or recommended dosage.

Chapter 5 - Toxic effects of heavy metals are a widespread concern in the modern toxicological studies. However, most studies dealing with the effects of heavy metals on invertebrates have addressed a few aspects of the organisms biology, and rarely attempt to analyse heavy metal effects at different levels of biological organization. The current trend in ecotoxicological studies proposes a use of a wide range of approaches involving several methods, each one focusing on different level of biological organization, to acquire a better understanding of noxius effects of heavy metals on marine invertebrate populations. The same metals may have contrasting effects on marine invertebrates depending on the biological level considered. As an example, here the authors present a review of several studies addressed to assess the ecological effects of sublethal concentrations of heavy metals on marine sponges. Responses at a physiological, molecular and early stages development levels differ according to the species and metal considered.

Sponges submitted to sublethal concentrations of heavy metals experience several behavioural and physiological responses such as changes in shape, metal accumulation, decreases in growth rates, reproduction output and filtration rates. Species differences in adult tolerance to heavy metals may be explained by differences in HSP expression. Induction of HSP production may allow some sponge species to endure heavy metal contamination. Studies on the effects of heavy metals on early development stages of sponges have shown contrasting results. While chronic expositions to heavy metals were harmful against adults, short expositions to similar concentrations of heavy metals resulted innocuous for larvae and juveniles of the same species. Thus, in order to acquire a general view of the effects of heavy metal on sponges, perhaps we should take into consideration all punctual responses given at different organization levels from organisms to molecules in order to avoid misinterpretation of the results.

Chapter 6 - Aquatic ecosystems have been heavily impacted by manufactured chemicals and other anthropogenic materials with dramatic effects on the microbial communities of these systems. However, few studies have examined how fine scale changes in environmental conditions resulting from natural or man-made forces can alter the structure, diversity, and function of aquatic microorganisms. The authors performed a number of studies that examined the influence of fine scale factors such as particle size or spatial location on microbial communities in freshwater systems. Bacterial communities associated with different sizes of particles in wetland sediment were characterized using 16S rRNA gene sequencing, which revealed fundamental differences in the structure of communities associated with very fine (0.063-0.25 mm diameter) and fine (0.25-1 mm) particles. Similarly, microbial activity was also influenced by particle size with different activities of lignin and cellulose degrading enzymes associated with different sized particles. Fine (cm) scale spatial variation in microbial activity was also seen on decomposing leaves in a stream, and spatial patterns in cellulose degrading enzymes suggest that the process of organic matter decomposition shows high spatial heterogeneity. These findings demonstrate that aquatic microbial communities vary both structurally and functionally over finer scales than are usually considered and these fine-scale interactions should be considered when assessing microbial communities for effects of or treatment and removal of environmental stresses.

Chapter 7 - The growing concern about the presence of genotoxic substances in water bodies is leading to an increase of the attempts to refine the assays for the evaluation of genetic damage. Performing cytogenetic tests on ecologically relevant species allows carrying out assays to detect early effects of environmental exposure to genotoxicants. Moreover, these tests have also been widely used in laboratory studies to investigate the genotoxic properties of several compounds. Among these assays, one of the most popular is the micronucleus test, because of its simplicity, sensitivity and reliability. Micronuclei are nuclear remains produced during mitosis (or meiosis) when a chromosome fragment or an entire chromosome fails to migrate with one of the two daughter nuclei formed. Basically, this assay consists in the observation of the variations of the frequencies of micronucleated cells. The micronucleus test has been performed in many aquatic species, both invertebrate (worms, sea urchins, mussels) and vertebrate (mainly fish and amphibians, but also mammals). The samples can be taken from any active tissue or alternatively by induction of cellular proliferation. As this test has not yet been validated, special attention should be paid to protocols, experimental design, confounding factors and inter-species differences.

Chapter 8 - Pneumonia with *Legionella spp.* presents a public health challenge especially because fatal outcomes still remain frequent. *Pseudomonas aeruginosa* is a significant source of hospital-acquired pneumonia and can also cause devastating chronic infections in compromised hosts, for example respiratory infections in cystic fibrosis patients. The aim of the authors studies was to describe the abundance and epidemiology of *Legionellaceae* in the man-made environment. In total, water systems of 70 different buildings in the German town of Göttingen (Lower Saxony) were examined for the presence of *Legionella* in two sampling cycles. Of these 22 (31%) had the bacterium in at least one water sample. *Legionella pneumophila* serogroups 1, 3, 4, 5 and 6 could be identified in the water samples. Most of the buildings were colonized solely by one distinct strain, as proven by PCR typing. Some buildings contained more than one PCR type or even more than one serogroup. Additionally the colonization of greenhouse misting systems with *Legionella spp.* and *Pseudomonas spp.* was studied in 20 different greenhouse misting systems located in Northern Germany. In total 80 water samples were collected. Each system was tested on two different occasions. Water was drawn at a central tap and at the outlet of spray nozzles. Sampled greenhouses were used to cultivate various plants and trees for commercial, recreational or scientific reasons, some of them in tropical conditions. *Legionella spp.* was detected in 10% of the systems (two systems), but only in low numbers. *Pseudomonas spp.* was recovered from 70% of the greenhouse watering systems (fourteen systems), occasionally at counts greater than 10,000 CFU/100 ml. Each colonized greenhouse had one or several individual strains of *Legionella* and *Pseudomonas* that could not be detected in any other system. This was demonstrated by a random amplified polymorphic DNA typing method. The possible health hazard caused by these water systems for both genera of bacteria is evaluated and discussed.

Chapter 9 - The aim of this work was to study Fe localization and Fe-dependent reactions in digestive glands of two sub-Antarctic limpets *Nacella (P.) magellanica* and *Nacella (P.) deaurata*. In spite of living in the same area, the difference in shore level location affects animal exposure to aerial and environmental conditions. The intertidal species (*N. (P.) magellanica*) is subjected to a variety of stress not experienced by *N. (P.) deaurata*; however *N. deaurata* would be exposed to seawater Fe for longer periods than *N. (P.) magellanica*. Total Fe content in digestive glands was 1.99 ± 0.03 and 0.99 ± 0.07 nmol/mg FW for *N. (P.) deaurata* and *N. (P.) magellanica*, respectively. Lipid radical content was assessed by electronic paramagnetic resonance (EPR), and was higher for *N. (P.) deaurata* than for *N. (P.) magellanica* (192 ± 31 and 122 ± 29 pmol/mg FW, respectively). The labile Fe pool (LIP), assessed by two different techniques, did not showed any difference between both species. Ferritin (Ft) was extracted from digestive glands, purified in a Sephacryl-S-300 column, and the protein was identified by SDS-PAGE. A significantly different Ft content in both species (22.9 and 6.17 ng/mg FW for *N. (P.) deaurata* and *N. (P.) magellanica*, respectively) was observed. The Fe content per molecule of Ft did not show any difference between the species. These results suggest that the Fe excess is efficiently stored in the Ft in *N. (P.) deaurata* since the LIP is not different between species even though *N. (P.) deaurata* showed a higher content of both Fe and Ft, than *N. (P.) magellanica*. However, other mechanisms could be operative in making Fe available for the catalysis of free radical reactions under stress conditions since lipid radical content was higher in *N. (P.) deaurata* as compared to *N. (P.) magellanica*.

In: Aquatic Toxicology Research Focus
Editor: Elias P. Svensson, pp. 1-29
ISBN: 978-1-60456-192-0

Chapter 1

OXIDATIVE STRESS AS A COMPONENT OF TRANSITION METAL TOXICITY IN FISH

Volodymyr I. Lushchak[1]
Department of Biochemistry, Precarpathian National University, 57 Shevchenko Str., Ivano-Frankivsk 76025, Ukraine

ABSTRACT

Reactive oxygen species (ROS) including superoxide anion ($O_2^{\bullet-}$), hydrogen peroxide (H_2O_2), hydroxyl radical ($^{\bullet}OH$) and their derivatives are continuously generated by different cell structures under aerobic conditions. They are catabolized by several specialized systems, which together with ROS generation create certain steady-state ROS concentrations. The situation when the balance between generation and degradation processes is disturbed in favor of the first, and the ROS steady-state concentrations are increased has been called "oxidative stress". Reactive oxygen species cause damage to almost all cellular constituents. In aquatic organisms oxidative stress may be developed naturally, for example, at fluctuations of environmental oxygen concentration, at temperature challenges and due to UV radiation, etc. In addition, many xenobiotics also increase production of ROS or compromise detoxification systems, or even both. For instance, some of them can either directly enter redox cycles or do it after biotransformation. Catabolism of certain xenobiotics by cytochrome system also can enhance ROS production as well as ions of metals with changeable valence. Inactivation of antioxidants or their exhaustion may also result in oxidative stress. Most studies to date show that toxicants can affect the redox balance via different mechanisms, which stimulate oxidative damage to cellular components, including proteins, lipids, nucleic acids etc. and exposure of cells to enhanced ROS levels leads to the dangerous intermediate production, mutagenesis or even death. The general description of ROS generation and development of oxidative stress in aquatic animals is analyzed with the focus on transition metal ions.

---

1 Fax: +38(0342) 714683; E-mail: lushchak@pu.if.ua.

I. Introduction

Many xenobiotics are capable to modulate metabolism of free radicals being either redox cycling compounds or transformed into them due to some sort of reactions. However, some compounds, particularly metal ions with non-changeable valence also can modify free radical processes. For example, calcium ions may substitute redox active bivalence iron or copper ions in binding sites and make them available for redox cycling (Halliwell and Gutteridge, 1989). Although over 90% of oxygen consumed by eukaryotic organisms is used by electron-transport chain, less than 5% is responsible for the production of free radicals and other active oxygen species. The last can interact virtually with any cellular component and dependently on the intensity of free radical generation and the efficiency of defense system may cause different effects. If the generation intensity overwhelms the antioxidant defenses, free radicals can damage cellular components with different consequences. Oxidative damages are related with many types of stresses, such as natural temperature and oxygen concentration fluctuations, solar and other irradiation, and anthropogenic pollution.

In the water environment organisms, particularly fish, are subjected to many interrelated factors which also fluctuate diurnally, seasonally and annually. The contaminations of anthropogenic origin to big extend by heavy metals and polychlorinated biphenyls are powerful modifiers of free radical processes, and in many cases the toxicity of mentioned above as well as many other xenobiotics is related to their capability to interfere with free radical processes.

This review presents a compilation of literature and authors' data on the modification of free radical processes in fish by xenobiotics with focus on most widely distributed and well studied ones, on transient metal ions.

II. General Aspects of Oxidative Stress Theory

Oxygen is absolutely essential for most living organisms and therefore they depend on its presence in the environment. Over 90% of oxygen consumed by organism is utilized by cytochrome oxidase – a terminal component of electron transport chain (Halliwell and Gutteridge, 1989). The enzyme catalyzes four-electron reduction of molecular oxygen to water. However, some electrons can „escape" electron transport chain and generate free radicals and other reactive species. A broad range of reactive species may be generated in living organisms, but here we will focus only on reactive oxygen species (ROS). The latter is a collective name for superoxide anion radical ($O_2^{\bullet-}$), hydrogen peroxide (H_2O_2), hydroxyl radical ($^{\bullet}OH$) and other their derivatives. Reactive oxygen species are oxygen-containing molecules, which are more chemically active than background state molecular oxygen. Superoxide anion is formed due to one-electron reduction of molecular oxygen:

$$O_2 + e^- \rightarrow O_2^{\bullet-} \quad (1)$$

Following one-electron reduction of $O_2^{\bullet-}$ with addition of two protons results in production of H_2O_2:

$$O_2^{\bullet -} + e^- + 2H^+ \rightarrow H_2O_2 \tag{2}$$

Further H_2O_2 molecule can accept one more electron and give rise to hydroxyl radical and hydroxyl anion:

$$H_2O_2 + e^- \rightarrow {}^{\bullet}OH + OH^- \tag{3}$$

Electrons for the reaction (3) may be provided by different donors for example $O_2^{\bullet -}$:

$$H_2O_2 + O_2^{\bullet -} \rightarrow O_2 + {}^{\bullet}OH + OH^- \tag{4}$$

Ions of metals with changeable valence also can donate electrons to H_2O_2 yielding hydroxyl radical and anion in Fenton reaction:

$$H_2O_2 + Fe^{2+} \rightarrow {}^{\bullet}OH + OH^- + Fe^{3+} \tag{5}$$

Oxidized iron ion (III) may be reduced via a reaction with $O_2^{\bullet -}$ to form molecular oxygen according to equation:

$$O_2^{\bullet -} + Fe^{3+} \rightarrow O_2 + Fe^{2+} \tag{6}$$

The net of reactions described by equations (5) and (6) gives the reaction (4) that for the first time was postulated by F. Haber and T. Weiss in 1934 and has become known later as Haber-Weiss reaction. This reaction can be catalyzed by ions of iron, copper, chromium, cobalt, titanium, vanadium and their complexes, etc.

Hydroxyl radical is more active and toxic than $O_2^{\bullet -}$ and H_2O_2. It can attack, for example, membrane unsaturated lipids and produce a series of radical reactions damaging the membranes (Esq. 7-10). Other oxidizing agents are also thought to promote lipid peroxidation *in vivo*, such as perferryl species and peroxynitrite ($ONOO^{\bullet -}$) (Valko et al., 2007):

$$LH + {}^{\bullet}OH \rightarrow L^{\bullet} + H_2O \tag{7}$$

$$L^{\bullet} + O_2 \rightarrow LOO^{\bullet} \tag{8}$$

$$LOO^{\bullet} + LH \rightarrow LOOH + L^{\bullet} \tag{9}$$

$$LOOH \rightarrow \text{products of decomposition,} \tag{10}$$

where LH and LOOH depicture lipid and lipid peroxides, respectively.

Reactions (7-10) demonstrate how lipid peroxidation is initiated by $^{\bullet}OH$ and further propagated. Hydroxyl radical and other ROS can oxidize in fact all cellular components, but currently scientists are mainly concentrated on most critically important compounds, namely proteins, nucleic acids and lipids.

Specific and non-specific defense systems prevent ROS generation, break the chain reactions, detoxify or hydrolyze the end products of tree radical attack of biomolecules. They collectively are named antioxidant systems. Antioxidants are grouped in two large groups in according to their molecular mass. There are low molecular mass radical scavengers like lipophilic compounds vitamin E (α-tocopherol), β-carotene, ubiquinone, vitamin A (rotenol) and water-soluble ones, vitamin C (ascorbic acid), catecholamines, uric acid, thiols, particularly reduced form of glutathione (GSH). The high molecular mass antioxidant system involves specific enzymes, such as superoxide dismutase (SOD; ECI.15.I.I), catalyzing the dismutation of superoxide (11), catalase (CAT; EC1.II.1.6) decomposing H_2O_2 (12), glutathione peroxides (GPX; EC1.II.1.9), which decompose organic and inorganic peroxides using GSH as a cosubstrate (13), and glutathione reductase (GR; EC 1.8.1.7) reducing GSSG to GSH using NADPH as the reducing power (14).

$$2O_2^{\bullet -} + 2H^+ \rightarrow H_2O_2 + O_2 \qquad (11)$$

$$2H_2O_2 \rightarrow 2H_2O + O_2 \qquad (12)$$

$$ROOH + 2GSH \rightarrow ROH + GSSG + H_2O \qquad (13)$$

$$GSSG + NADPH + H^+ \rightarrow 2GSH + NADP^+ \qquad (14)$$

Under certain conditions sulfur-containing amino acid residues of proteins, oxidized by ROS, may be reduced to their initial forms by special enzymes. For example, dithiols produced at cysteine residue oxidation may be reduced by thioredoxin (EC 1.20.4.1) in according to reaction (15), while methionine sulfoxide reductase reduces oxidized form of methionine – methionine sulfoxide (16). Both these reactions use NADPH as a donor of reductive eqivalent.

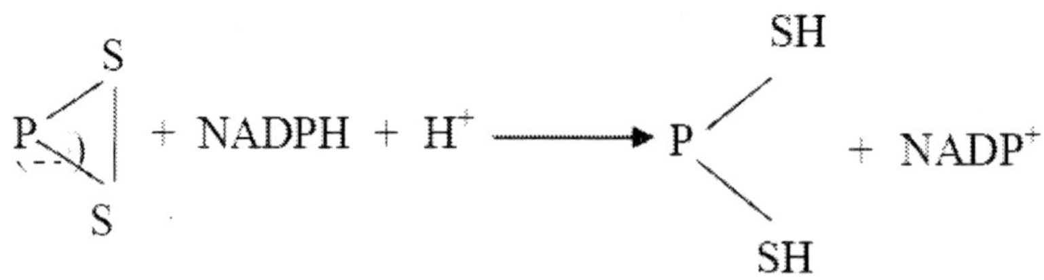

(15)

$$\text{Methionine sulfoxide} + NADPH + H^+ \rightarrow \text{Methionine} + NADP^+ \qquad (16)$$

In normally functioning cell the processes of generation and degradation of ROS usually are well balanced. Therefore the steady-state ROS concentrations are extremely low with small fluctuations. For example it is known that the concentration of H_2O_2 is usually in range 10^{-9}-10^{-10} M, while $O_2^{\bullet-}$ concentration is even lower – 10^{-10}-10^{-11} M (Halliwell and Gutteridge, 1989). However, under certain conditions the balance between production and degradation of ROS may be shifted in favour of the first, which can result in increase of steady-state ROS concentration. i.e. oxidative stress (Sies, 1985; Lushchak, 2001). Increased ROS generation, reduced degradation or both these processes together are responsible for oxidative stress development. Oxidative stress may either result in development of many animal and human pathologies, or accompanies them. Injuries produced by ROS are mainly concentrated at the places, where ions of metals with changeable valence, primary ion and copper, are localized. Reactive oxygen species may attack in fact all cellular components as above mentioned. For example, ROS interaction with proteins results in the modification of amino acid residues, including oxidation of cysteine, methionine, tyrosine, phenylalanine, tryptophan, etc. (Stadtman, Levine, 2000; Dunlop et al., 2002; Requena at al., 2003; Lushchak, 2007). The attack by ROS on nucleic acids gives rise to a big set of products which can be repaired in some cases, but often leads to irreparable changes resulting in mutations (Demple 1990; Demple and Sung, 2005).

Polyunsaturated fatty acids are especially sensitive to ROS attack resulting in the initiation of lipid peroxidation and giving a wide range of products, like lipid peroxides, different aldehydes, acids, lysolipids, etc. (Vladimirow, 2002). Since ROS are very harmful for cell function, there is a rather complicated multilevel system that serves for protection against their toxic effects. The systems of protection against ROS now are categorized into three groups: 1 – prevention of ROS generation, 2 – break of free radical chain and detoxification of radicals by quenchers and antioxidant enzymes, and 3 – repair of damaged elements (Lushchak, 2002; 2003).

Binding of changeable valence metal ions by specific and unspecific proteins such as ferritin, transferrin, albumin and others is one more important protection mechanism (Sies, 1993). Low molecular mass antioxidants like tocopherol, ascorbic and uric acids realize their antioxidant potential mainly by breaking free radical chain reactions and stopping their propagation (Halliwell and Gutteridge, 1989). Glutathione directly quenches hydroxyl radicals and singlet oxygen and/or serves as a cofactor for some enzymes, and is also involved in regeneration of reduced vitamin C and tocophyryl radical (Sies, 1993). The described above material may be summarized in general schemes of ROS generation and detoxication (Figure. 1).

The cell consuming molecular oxygen, use it mainly for the energy production via operation of electron transport chain which transfers electrons onto oxygen molecule at the level of cytochrome oxidase. This reaction does not produce ROS, but electrons may leak at other parts of electron transport chain providing one-electron reducing either molecular oxygen or ROS. Superoxide anion radical may either directly interact with cell constituents or be reduced to H_2O_2. The hydrogen peroxide dismutation is substantially accelerated by catalase. The formed hydrogen peroxide either interacts with different compounds or is reduced via several routs. Nonenzymatic reduction of H_2O_2 provided by $O_2^{\bullet-}$ and ions of transition metals gives rise to $^{\bullet}OH$ and OH^-. There are several possibilities to prevent realization of the danger scenario with $^{\bullet}OH$ production. That is mainly enzymatic degradation of H_2O_2 by different peroxides particularly by catalases and glutathione-dependent peroxidases. The first produce water and molecular oxygen, while the latter produce water for expenses of GSH. Formed oxidized glutathione is reduced by glutathione reductase for the expenses of NADPH. It should also be noted that any ROS may also directly or indirectly interact also with low molecular mass antioxidants, which may break development ROS generation. The balance between these and other processes which ROS are involved in is under fine cellular control, but under some circumstances the steady-state ROS concentration may be increased causing oxidative stress and leading to poorly predictable consequences.

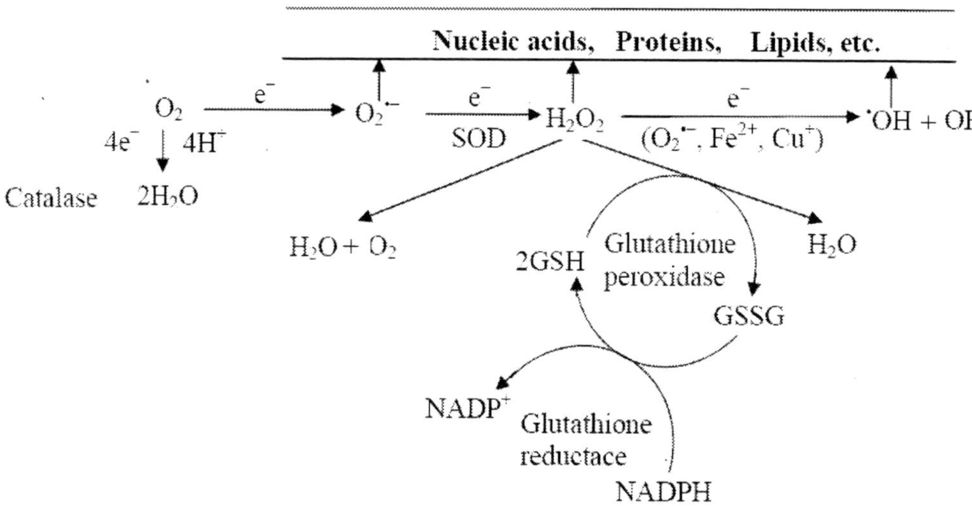

Figure 1. The metabolism of reactive oxygen species and involvement of transition metal ions in it.

III. Generation of ROS in the Cell

There are two principal sources of superoxide radical in eukaryotic cells – respiratory chain of mitochondria and NADPH-oxidase complex of the cytoplasmic membrane. In mitochondria, production of $O_2^{\bullet-}$ occurs at two sites – at the level of NADH dehydrogenase and at ubiquinone-cytochrome *b* part. Accepting the electron ubiquinone is converted into ubisemiquinone radical. This radical can donate the electron either to cytochrome C (normal pathway) or molecular oxygen (superoxide anion producing pathway). Under normal conditions over 90% of cellular $O_2^{\bullet-}$ is produced by mitochondria. The inhibition of electron transport chain below ubisemiquinone results in reduction of electron carriers and therefore increases $O_2^{\bullet-}$ production owing to switching the electron flow directly to molecular oxygen. It was shown that cyanide and antimycin A dramatically increased superoxide production by mitochondria (Vladimirov, 2002). For the same reason, the loss of cytochrome *c* by mitochondria also increases the superoxide anion production (Vladimirov, 2002). Several more experimentally found events leading to change in ROS production by mitochondria also should be listed: reduced cytochrome oxidase activity in aged organisms increased ROS production; decreased ROS production in the presence of protonophores; increased ROS generation at higher ambient O_2 concentrations; and the activation of $O_2^{\bullet-}$ release by mitochondria in the presence of calcium ions (Sheehan et al., 1997; Vladimirov, 2002).

Plasma membrane possesses a system directly adding electron to molecular oxygen yielding $O_2^{\bullet-}$. This system is highly expressed in immune cells namely neutrophils consuming little oxygen since they rely mainly on glycolysis for ATP production and are rich in stored glycogen. On the other hand, macrophages possess mitochondria and produce most ATP via oxidative phosphorylation and therefore consume more oxygen than neutrophils. At phagocytosis both cell types show great increase in oxygen consumption, which has been called "respiratory burst". But sharply increased oxygen consumption it not inhibited by cyanide (therefore called as cyanide insensitive), an inhibitor of mitochondrial electron transport chain. During the respiratory burst cells also consumes more glucose via stimulation of pentose phosphate pathway. It was shown, that the oxygen uptake activation was due to the activation of complex enzyme associated with plasma membrane. The system oxidizes NADPH provided by the pentose phosphate pathway in according to equation:

$$NADPH + 2O_2 \rightarrow NADP^+ + H^+ + 2O_2^{\bullet-} \quad (17)$$

The NADPH complex is arranged vectorially in the membrane and therefore electrons pass across it from the NADPH-oxidizing intracellular site to the external O_2-reducing site. The movement of electrons across the plasma membrane seems to be accompanied by outward movement of protons through proton-conducting channels in the membrane. Physically, the NADPH-oxidase complex is a FAD-containing flavoprotein. It is believed that flavoprotein is reduced by NADPH and transfers electrons to cytochrome *b*, which donate them further to O_2 molecule. The generation of ROS, as well as the active species of nitrogen and halogens, is responsible for bacteria killing by phagocytic cells.

The endoplasmic reticulum of many animal tissues contains cytochromes collectively known as cytochrome P-450. They were named in this manner because their reduced forms in complex with carbon monoxide absorb light strongly at 450 nm. Cytochrome P-450 oxidizes a wide range of substrates using molecular oxygen. It adds one atom from oxygen molecule to substrate, while other forms water. Therefore this reaction has been called mono-oxygenase or mixed-function oxidase (Vladimirov, 2002). The overall reaction catalyzed by cytochrome P-450 can by represented by the following equation:

$$AH + O_2 + RH_2 \rightarrow A\text{-}OH + R + H_2O, \quad (18)$$

where RH_2 is a reducing agent and AH is a substrate. In the liver cells, electrons required by the cytochrome P-450 system are donated by NADPH via flavoprotein enzyme NADPH-cytochrome P-450 reductase. The flavoprotein enzyme transfers electrons from NADPH to adrenodoxin. The reduced form of adrenodoxin, putiredoxin, and NADPH-cytochrome P-450 reductase from liver can „leak" electrons onto molecular oxygen, reducing it to give $O_2^{\bullet-}$. In addition, some oxygenated intermediates of cytochrome P-450 can decompose in a minor side reactions and release $O_2^{\bullet-}$. Substrates for the cytochrome P-450 system include sterol hormones, insecticides such as heptachlor and aldrin, hydrocarbons such as benzpyrene, and drugs such as phenacetin, amphetamine, paracetamol, etc. Usually the products of reaction catalyzed by cytochrome P-450 are less toxic than initial material, but it is not always. For example, hydroxylated products of paracetamol and cancerogenic hydrocarbons such as benzpyrene actually cause cellular damage stronger, than their precursors (Senthilnathan et al., 2006). It is interesting to note, that there is so-called ethanol-inducible cytochrome P-450, efficiently oxidizing ethanol, other alcohols and acetone. Finally, we should recognize that some part of used by cytochrome P-450 oxygen is released as $O_2^{\bullet -}$, i.e. this system can produce ROS as by-products.

IV. Metal Ion Uptake by Fish

Different metal ions are essential for the health of many organisms. However, their excess may be toxic and often their toxicity is related to generation of ROS. Therefore, a fine balance between metal deficiency and surplus is maintained via the regulation of absorption and excretion. Fish have two ways of metal acquision – from a diet and from a water. Therefore there are at least two main organs responsible for ion absorption – the gills and intestine. Although the ion transporters have been identified and extensively studied in animals, plants and yeast, the information on fish is scarce. Since there is no clear description of mechanisms of ion uptake in fish, we will give a hypothetical view of this process using homology with transport systems in other organisms, particularly in yeast (Gospodarev and Lushchak, 2005) like it was done by Bury and colleagues (2003). Here we will focus only on several ions, predominantly on iron because of its importance and our own experience with the element, particularly in yeast (Gospodaryov and Lushchak, 2005; Gospodaryov et al., 2005) and fish (Bagnyukova et al., 2006).

Being an essential element iron is involved in many cellular processes; in excess it can be detrimental for health. It easy enters redox processes and therefore is responsible for ROS generation. In addition, excess of waterborne iron may be toxic for fish due to the formation of iron „flocks" on the gills and intestine, resulting in their clogging (Bury at al., 2003; Bagnyukova et al., 2006). Although iron is one of the most abundant elements on Earth, in aerobic environments it is predominantly in ferric (hydro/oxides) forms, which are relatively insoluble at neutral pH. Therefore the ionic ferric (Fe^{3+}) concentrations are exceedingly low. Many unicellular aquatic organisms have evolved specialized transport systems to obtain sufficient iron to meet metabolic demands. They excrete extracellular chelators of iron, known as siderophores possessing exceedingly high affinities for iron. The latter forms complexes with iron ions and is absorbed by cells (Gospodarev and Lushchak, 2005). In the cell the complexes dissociate and iron-free siderophores are excreted in environment to repeat the cycle. It is interesting to note, that some microorganisms, which do not produce sideropnores, may use ones produced by other microbes (Gospodarev and Lushchak, 2005).

The mechanisms of iron, copper and zink uptake in teleost fish have been analyzed in details by Bury and colleagues (2003) and therefore here we will demonstrate only general aspects of iron exchange (Figure 2). In environment, iron ions predominantly exist in Fe^{3+} form, which cannot be directly absorbed by gills. However, fist it should enter mucus layer either of gills or intestine. Later Fe^{3+} must be reduced to ferrous ion Fe^{2+} by apical membrane-bound ferric reductase. The formed ferrous ion enter the cell either at the ferric reductase place or via Fe^{2+}/H^{+}-symporter using inward directed proton gradient. Into the cell iron can be used or kept in both ferrous and ferric forms. Basolateral Fe^{2+} export occurs via an iron regulated transporter also known as ferroporin. The latter is linked to a membrane-bound copper-containing oxidase (haphaestin), which oxidizes Fe^{2+} to Fe^{3+}. This process also may be due to nonenzymatic oxidation. Ferric ions are bound in the blood to transfer and transported to places of need.

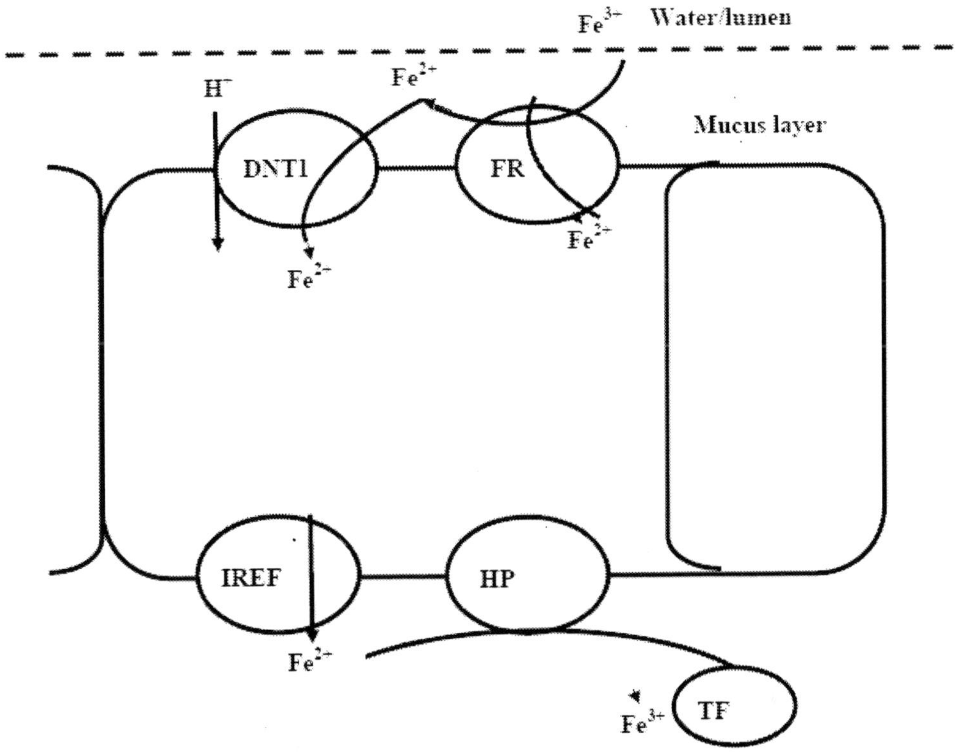

Figure 2. Hypothetical scheme of cellular iron uptake and excretion in fish – combining data from gill and intestine. Abbreviations: FR – ferric reductase; DMT1 – Fe^{2+}/H^{+} symporter; IREG – iron regulated transporter, HP – copper-containing oxidase (hephaestin); TF – transferrin. (Modified from Bury et al., 2003).

V. Transition Metal Ions and Free Radical Processes in Fish

There is a lot information on generation of oxidative stress in fish, but it is not easy to find systematically analyzed one. Therefore here we will go step-by-step via different, actually most important and studied elements, starting from general description of individual element going through their possible valences, redox properties to effects on living organisms and finally to their effects on fish. The described elements include iron, copper, chromium, mercury and arsenic.

V.1. Iron

The electronic structure of iron results in its capacity to drive one-electron reactions and together with relatively high content of this element in cells it seems to be one of the major components in the production and metabolism of free radicals in biological systems. In the latter iron commonly is found in three oxidation states: II, III and rare in state IV. Under physiological conditions, particularly at neutral pH values Fe^{3+} precipitates in a form of oxyhydroxide polymers, while Fe^{2+} is soluble. However Fe^{2+} is unstable in physiological solutions and tends to be easily oxidized, particularly by molecular oxygen yielding superoxide anion.

Interaction of iron ions with activated form of oxygen gives a large array of highly active compounds, particularly hydroxyl radical ($^{\bullet}OH$) – reactions (4, 6) (Valko et al., 2007). The overall iron ions play a role of catalyzers and are not consumed or transformed during described reactions (4,6) and 19:

$$LOOH + Fe^{2+} \rightarrow Fe^{3+} + LO^{-} + {}^{\bullet}OH, \tag{19}$$

where LOOH depictures lipid peroxides.

The formed hydroxyl radical can react at diffusion-limited rates with virtually all cellular constituents, including lipids, proteins, nucleic acids, carbohydrates, etc. Therefore isolation of iron ions from cellular components is one of the very efficient ways to prevent the generation of free radicals in biological systems. Usually it involves their chelating by ligands via formation of coordinative complexes with oxygen, nitrogen and sulfur and this binding limits iron ability to undergo redox cycles.

The metabolism of iron in eukaryotes is rather well studied (Aisen et al., 2001; Ponka, 1999). Virtually all investigated organisms depend on iron for survival. Iron-dependent life must deal with the paradoxal hazards of iron deficiency and iron overload, each with serious or fatal consequences. Last decade many aspects of iron metabolism were deciphered particularly because of use model organism budding yeast *Saccharomyces* cerevisiae (Van Ho et al., 2002; Kaplan et al., 2006). Genetic approaches uncovered long-sought hemochromatosis gene and connection between iron metabolism and Friedreich's ataxia (Aisen et al., 2001). Some information on iron requirement for fish may be found in the literature. For example, Watanabe and colleagues (1997) gave nutritional requirements of fish for dry diet. It is widely accepted that the iron content of fish is in general, considerably lower than that of other vertebrates, but the precise daily iron requirements for fish at present are unknown (Bury at al., 2003). It is assumed that iron metabolism and function in teleost fish is similar to other vertebrates. Fish obtain iron from water by uptake across the gill epithelium and by intestinal uptake from food (Bury et al., 2003). Iron metabolism in rainbow trout at normal and iron deficient diets have been studied in details (Walker and Fromm, 1976; Carriquiriborde et. al, 2004). Although iron is an essential element undergoing redox cycle and involved in initiation and propagation of free radical processes, the information of free radical processes induced by this element in fish is scarce.

In a series of experiments African catfish *Clarias gariepinus* during five weeks were fed by normal and iron enriched diets (Baker et al., 1997). Ingestion of the higher dietary iron ration suppressed fish growth indicating that the metal was supplied at sublethal, but toxic levels. The tissue iron concentrations were unaffected by dietary regimen. However, concentrations of malonic dialdehyde, a product of free radical lipid peroxidation, in liver and heart increased concomitant with dietary ion dose. At the same time, fat-soluble antioxidant α-tocopherol (vitamin E) was significantly depleted in liver of fish fed by high iron diet. The dynamic of the last two indices, malonic dialdehyde and α-tocopherol, demonstrated the development of oxidative stress induced by high iron level in food (Baker et al., 1997).

The effects of waterborn iron on free radical processes in goldfish *Carassius auratus* liver and kidney were studied in our laboratory (Bagnyukova et al., 2006). For experimentation fish were exposed to 20 and 500 μM of ferrous sulphate during 7 days. The treatment increased the levels of protein carbonyl groups, a marker of oxidative modification of proteins, but reduced concentrations of lipid peroxides. At the same time, concentrations of thiobarbituric acid reactive substances, a measure of the end products of lipid peroxidation, were increased in liver and kidney of fish treated with 500 μM of iron sulphate. Described above changes in markers of oxidative damage demonstrate a development of oxidative stress in goldfish tissues. Fish exposure to high environmental iron concentrations did not affect the activity of superoxide dismutase, but reduced the activity of catalase. The activity of glutathione-S-transferase was decreased at high iron ion concentrations, but glutathione reductase activity was reduced only in kidney of goldfish treated by 20 μM ferrous sulphate. This treatment did not change glucose-6-phosphate activities in liver, but again reduced it in kidney of animals exposed to 20 μM of ferric ions. A strong positive correlation between lipid peroxidation products from one side and the activities of catalase in the liver glutathione reductase in kidney from the other side indicated the possible upregulation of the enzymes by these products. We concluded on the existence of coordinated response of antioxidant systems on oxidative stress induced by high concentrations of environmental iron ions (Bagnyukova at al., 2006).

V.2. Copper

Copper (cuprum) is essential element for the function of most living organisms. Cupric ion (Cu^{2+}) has only 9 electrons in the *d*-orbitals with one unpaired electron. It represents the most stable oxidation state of copper and because of one unpaired electron Cu^{2+} is paramagnetic. The possibility to have also electronic state Cu^{+} creates basis for operation of Cu^{2+}/Cu^{+} rebox pair which helps to catalyze the movement of electrons in biological systems because of high redox potential. Cupric ions in the presence of biological reductants such as GSH or ascorbic acid can be reduced to cuprous ion (Cu^{+}) which is capable of catalyzing the formation of $^{\bullet}OH$ through decomposition of hydrogen peroxide via Fenton reaction (Valko et al., 2005):

$$Cu^{+} + H_2O_2 \rightarrow Cu^{2+} + {}^{\bullet}OH + OH^{-} \quad (20)$$

Analogously to iron, cupric ions may be reduced to cuprous ions by the second $O_2^{\bullet-}$ molecule:

$$Cu^{2+} + O_2^{\bullet-} \rightleftarrows Cu^{+} + O_2 \quad (21)$$

Therefore copper ions are prone to participate in the formation of ROS in biological systems and it is widely believed that this particular property is at least partially responsible for copper-induced cellular toxicity. For example, copper is capable for induction of DNA strand breaks and oxidation of bases via oxygen free radicals. The limit of „free" pool of copper is for less than a single free ion per cell (Rae et al., 1994). Therefore, one should bear in mind this significant overcapacity for copper chelation in the cell. Copper ions may be also responsible for oxidative modification of lipids and proteins (Kapiotis et al., 2005). From the other side, copper ions may be involved in antioxidant defense mechanisms. For example, being a component of Cu,Zn-SOD it catalyses the dismutation of $O_2^{\bullet-}$ and redox properties of copper ions are critical for this activity.

Since the copper is an integral part of many proteins needed for neurological functioning, its role in oxidant-mediated stress in neurodegenerative diseases was intensively studied. The role of copper has been implicated in neurodegenerative disorders such as Alzheimer's disease, Parkinson's disease and amyotrophic lateral sclerosis (Shcherbatykh and Carpenter, 2007; Rotilio and Stella 2006; Banci et al., 2007; Valko et al., 2005).

Copper homeostasis in fish entails on regulated uptake, transport and excretion similarly to mammals. However, there is some specificity of these processes in fish related with possibility of bronchial uptake along with intestinal. Both ways are efficient and their efficiency depends on many factors and is highly regulated process. For example, dietary copper pre-exposure downregulated bronchial uptake of waterborne one, where as deprivation of copper upregulated bronchial uptake (Kamunde et al., 2002). In rainbow trout gills two systems of copper transport have been described – the putative high affinity with low capacity and low affinity high capacity ones. Both of them are regulated by organism needs, water and diet concentrations of copper. Kamunde and colleagues (2002) studying rainbow trout suggested a quantitatively greater role for the dietary than for the waterborne rout of copper uptake, a key role for the gill in its homeostasis, and important roles for the liver and gut in the normal copper metabolisms in fish.

Copper toxicity to many fish species was studied extensively. For example, in rainbow trout 730 mg Cu/kg diet was toxic and characterized by reduced growth, increased feed:gain ratios, food refusal and elevated liver copper levels (Lanno et al., 1985). Increased levels of dietary copper did not affect significantly ascorbic acid content in kidney and liver, plasma copper, plasma glucose, hemoglobin or hematocrit as well as hepatosomatic index of trout. However, both copper and iron levels in the liver increased in relation to increased levels of dietary copper, which was interpreted as an interaction between copper and iron metabolism in the trout (Lanno et al., 1985). Trout exposure to subletal copper concentrations during a week did not induce respiratory stress, while common (*Cyprinus carpio*) and gibel (*Carassius auratus gibelio*) carps decreased considerably arterial oxygen tension (De Boeck et al., 2007). At the same time, arterial carbon dioxide tension was elevated meaning that gas exchange was compromised in these both cyprinid species. Gibel and common carps showed a significant decrease in plasma pH and increase in plasma lactate concentrations demonstrating stimulation of glycolysis under copper intoxication. Gibel carp also increased the production of ethanol as the end product of the alternative anaerobic pathway. Sodium turnover was supposed to be a key factor in predicting copper toxicity (De Boeck et al.,

2007). The activation of glycolysis under treatment with high copper concentrations may be due to mitochondria dysfunction. It has been known that at copper overload, the mitochondrion is a major target for toxicity, with oxidative mitochondrial membrane damage and poisoning of enzymes of the tricarboxylic acid cycle and energy metabolism as potential targets (Mehta et al., 2006). Water salinity may affect copper toxicity like it was found with euryhaline fish *Fundulus heteroclitus* (Blanchard and Grosell, 2006).

The chronic dietary exposure to copper also affected the growth and metal composition of the grey mullet *Chelon labrasus* (Baker et al., 1998). The high copper diet also reduced food intake. Hepatic α-tocopherol concentrations were 63% lower, while malonedialdehyde increased by 304% in mullet fed a high copper diet for 67 days. Therefore the authors concluded that the oral copper exposure induced oxidative stress in grey mullet (Baker et al., 1998).

The oxidative stress was demonstrated to be induced by both dietary and waterborne exposure to high copper, or even at injection. In gilthead seabrams *Sparus aurata* injection with $CuCl_2$ increased concentrations of thiobarbituric reactive substances (Pedrajas et al., 1995). The treatment increased the specific activity of superoxide dismutase and resulted in appearance of two new Cu,Zn-SOD isoforms. The new SOD isoforms were generated also *in vitro* by incubation of cell-free extracts with systems generating superoxide anion or hydrogen peroxide. The authors supposed that new SOD isforms found in fish treated with copper were a result of oxidative modification of inital forms (Pedrajas et al., 1995). Therefore it was concluded that copper injection induced oxidation of both lipids and proteins in gilthead seabram tissues.

The 40 day exposure of goldfish *Carassius auratus* to 0.005-0.025 mg/l copper reduced catalase, 0.05-0.05 – selenium dependent glutathione peroxidase activities, while 0.0025-0.01 mg/l increased glutathione-5-transferase activities (Liu et al., 2006). The modification of activities of antioxidant enzymes also demonstrates that copper induced oxidative stress in goldfish tissues. In warm water fish African walking catfish *Clarias gariepinus* dietary copper exposure also elevated copper concentration in the intestine, liver and gills in tissue-specific manner (Hoyle et al., 2007). The fish exposure to copper decreased also food intake and specific growth rate. At the same time, the treatment with copper significantly increased TBARS concentrations in the gill and intestine, while total glutathione content in the intestine was doubled. The liver also showed glycogen depletion consistent with reduced food intake, but not pathological changes were found in the gills, liver or intestine (Hoyle et al., 2007).

Hansen and colleagues (2006a,b) investigated the effects of water contamination on the activities and steady-state concentrations of mRNA of certain antioxidant enzymes in brown trout tissues. The general conclusions, which may be drawn from their studies are: (*i*) fish exposure to metal ions, particularly to copper, enhances the activities of primary (SOD, catalase, glutathione peroxidase) and secondary (glutathione reductase, metallothionein) enzymes/proteins; (*ii*) the level of mRNA do not always correspond to respective protein level and (*iii*) metallothioneins do not necessary are up-regulated by the addition of metal ions like copper.

Because copper is a component of several antioxidant systems, its deficiency may be expected to induce oxidative stress also. For example, copper ions are integral parts of cytochrome oxidase and Cu,Zn-containing SOD. In such situation, copper ions are clearly antioxidants or elements of the electron transport chain. However, either reduced uptake or disrupted intracellular metabolism theoretically may substantially increase a steady-state ROS concentration due to copper ion cycling. The enhancement of copper concentration results in saturation of cell with this micronutrient and antioxidant systems which rely on this element operate efficiently. Therefore, the concentration of free intracellular copper is kept at loosely low level. Further increase of intracellular copper pool may saturate all possible chelators and if it is overwhelmed, new their molecules are synthesized to prevent the appearance of free copper ions. Metalothioneins and chaperons play a critical role in binding of copper when the ions are in excess. Under these conditions the production and degradation of ROS are well balanced and the only negligible oxidative stress takes place. The situation is critically changed when the cell is not able to bind the excess of copper ions. In this case, they may enter Haber-Weiss reaction producing harmful hydroxyl radical resulting in stimulation of oxidative processes induced by free radicals. The described above relationship between the intensity of free radical processes and intracellular concentration is depictured at Figure 3.

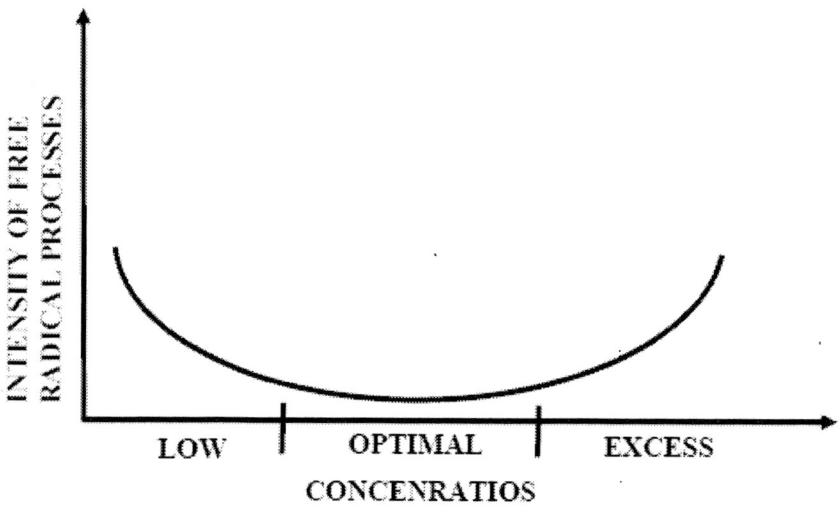

Figure 3. The relationship between the intensity of free radical processes and intracellular concentration of copper ions.

V.3. Chromium

Chromium occurs in the workplace predominantly in two valence states – hexavalent (Cr^{6+}) and trivalent (Cr^{3+}). Hexavalent chromium compounds are used widely in diverse industries and trivalent chromium salts, for example, chromium picolynate, chromium chloride and niacin-bound chromium are used as micronutrients and dietary supplements. Like other metals, chromium may be toxic and does not have biodegradable properties, remaining in ecosystems. The direct or indirect bioaccumulation in organisms in contaminated waters may be significant and therefore it may affect not only individual organisms, but also the food chains.

Chromium exists in several states of oxidation, ranging from +2 to +6, but its forms +3 and +6 are the most stable in the environment and biologically important. Being an element with changeable valence (II, III, IV, V and VI) it can enter Haber-Weiss-type reaction resulting in generation of ${}^{\bullet}OH$ radical and at least several types of this reaction may be found in the literature (Halliwell and Gutteridge, 1989; Shi et al, 1990; Valko et al., 2005).

$$Cr^{6+} + H_2O_2 \xrightarrow{H_2O_2} Cr_2O_2^{3-} \xrightarrow[\text{reductant}]{\text{cellular}} Cr^{5+} + {}^{\bullet}OH \quad (22)$$

$$Cr^{6+} + H_2O_2 \longrightarrow CrO_8^{3-} \xrightarrow[\text{reductant}]{\text{cellular}} Cr^{5+}\text{ complex} \xrightarrow{H_2O_2} {}^{\bullet}OH \quad (23)$$

$$Cr^{6+} \xrightarrow[\text{diols}]{\text{GSH, enzymes}} Cr^{5+}\text{ complex} \xrightarrow{H_2O_2} {}^{\bullet}OH \quad (24)$$

The given above reactions may be presented in another form:

$$Cr^{6+} + O_2^{\bullet-} \longrightarrow Cr^{5+} + O_2 \quad (25)$$

$$Cr^{5+} + H_2O_2 \longrightarrow Cr^{4+} + {}^{\bullet}OH + OH^{-} \quad (26)$$

and overall

$$O_2^{\bullet-} + H_2O_2 \xrightarrow{Cr^{6+}/Cr^{5+}} {}^{\bullet}OH + O_2 + {}^{\bullet}OH \quad (27)$$

Two aspects should be noted: (*i*) cellular reductions are needed for generation of ${}^{\bullet}OH$ and reactions 23-25 (*ii*) chromium may play a role of catalyst being able to enter reversible oxidation (reactions 26-28). Therefore it is commonly accepted that biological effects of chromium, at least partially, are connected with ROS generation.

Whereas trivalent chromium is an essential trace element, hexavalent chromium is non-essential, toxic, and corrosive and causes allergic reactions, damage to liver and kidneys, besides showing carcinogenic effects in plants (Leonard et al., 2004). Chromium absorption by animal organisms depends on its oxidative state: Cr^{6+} being gastrically absorbed between 2 and 6%, and Cr^{3+} around 1%. In blood stream chromium binds to plasma proteins and is excreted mainly in urine (Tagliari et al., 2004; Bagchi et al, 2002). Probably $K_2Cr_2O_7$ is the most widely distributed human-born form of chromium. The chromate ion (CrO_4^{2-}), is a dominant Cr^{6+} form in aqueous solutions. It quickly penetrates the cell membrane via non-specific anion channels and is reduced to trivalent form Cr^{3+}. The reduction of hexavalent chromium to Cr^{5+} and Cr^{3+} was described in Chinese hamster V-79 cells (Ueno et al., 1995) and in isolated rat hepatocytes (Gunaratnam and Grant, 2001). In the last case, both DT-diaphorose and cytochrome P-450 played only a minor role in detoxifying Cr^{6+} and/or its metabolites. It appeared that GSH acts via non-enzymatic reduction, reducing Cr^{6+} to more toxic form. Glutathione reductase appeared to play an important role in the enzymatic reduction of Cr^{6+} because inhibition of this enzyme by carmustine markedly protected the cells from cytotoxicity (Gunaratnam and Grant, 2001).

In rat hepatocytes Cr^{6+} increased ROS formation, lipid peroxidation, decreased mitochondrial membrane potential and resulted in rupture of lysosomal membranes (Pourahmad and O'Brien, 2001). In the same cells, the mitochondria were found to be a major contributor to endogenous ROS formation under chromium treatment (Pourahmadet al., 2003). In turn, ROS caused a loss of lysosomal membrane integrity and the release of lysosomal proteases and phospholipases also contributed to hepatocyte cytoxicity. Authors concluded that lysosomes are also sites of cytotoxic ROS formation with Cr^{6+} (Pourahmadet al., 2003). *In vitro* studies with plant and animal mitochondria clearly demonstrated that Cr^{6+} induced perturbations of mitochondrial bioenergetics compromising energy-dependent biochemical processes, which also may contribute to the basal mechanisms unreeling its toxic effects in plant and animal cells (Fernandes et al., 2002). Chromium may stimulate not only oxidation of cellular lipids and proteins, but DNA as well. For example, *in vitro* Lloyd and colleagues (1997) found that Cr^{6+} induced intra-strand crosslinks and strand breaks in salmon sperm DNA. The presence of potassium chromate (K_2CrO_4) in drinking water (100 and 200 p.p.m. chromium) induced formation of DNA-protein crosslinking in liver, but not in lymphocytes of rats (Coogan et al., 1991). Using the alkaline single cell gel electrophoresis (Commet assay) Polish team studied the effects of hexavalent chromium on the DNA damage and repair in human lymphocytes (Blasiak and Kowalik, 2000). Potassium dichromate induced DNA damage in the lymphocytes and effects were time- and dose-dependent. Cr^{3+} caused greater DNA migration than Cr^{6+}. Although the Commet assay did not disclose the mechanisms in the DNA-damaging activity of Cr^{3+}, the authors speculated that its binding to cellular ligands may play certain role in its genotoxicity (Blasiak and Kowalik, 2000). The formation of DNA single strand breaks in rat hepatocytes treated with $K_2Cr_2O_7$ was inhibited by deferrioxamine, an iron chelator (Susa et al., 1997). Pretreatment of rat hepatocytes with α-tocopherolsuccinate (vitamin E derivative) prior to exposure to $K_2Cr_2O_7$ resulted in marked decrease of Cr^{6+} -induced cytotoxicity, as evaluated by the leakage of lactate dehydrogenase, without affecting cellular uptake and the subcellular distribution of chromium (Susa et al., 1996). Another antioxidant taurine partially reduced Cr^{6+}-induced oxidative stress and chromium accumulation in mice kidney (Bosgelmez and Guvendik, 2004). The data on protective role of antioxidants once more confirm the ROS involvement in cellular damage by chromium and beneficial potential of antioxidants to prevent this damage. Most results of chromium effects on living organisms have been received on mammals and their effects on fish are very limited. However, the question can not be ignored in aquatic organisms, especially in fish. They may be subjected to chromium compounds not directly because they often occupy the top position in aquatic food chains.

As it was mentioned above, chromium has both beneficial and deleterious effects in organisms being essential trace element involved in regulation of broad array of biological processes, particularly in glucose metabolism. In experiments with guppies *Poecilia reticulata* Perez-Benito (2006) found that low concentrations ($<10^{-4}$ M) of Cr^{6+} increased the maximum lifespan in both mails and females, while vanadium (VI) did it only in mails. The maximum lifespan showed a strong, positive correlation with the concentration of chromate for males, and a weaker one for females. Author suggested the existence of a chemical-hormesis phenomenon that might be related to sexual-genre variability. Both the toxicity and the chemical-hormetic effects of chromate were substantially decreased when it was used in combination with antioxidant D-mannitol. The latter might indicate the ROS involvement in described effects of chromate (Perez-Benito, 2006). Chromium sulfate at sublethal concentrations increased plasma sodium level and decreased plasma chloride level in common carp *Cyprinus carpio* (Subashini et al., 2005). The activity of Na,K-ATPase of red blood cells was inhibited by this treatment. Authors concluded that ion transport system may be used as a valuable biomarker of pollutant exposure (Subashini et al., 2005). In other study with coastal teleost fish *Periophthalmus dipes,* potassium chromate effect on the ion-dependent ATPases of gills, kidney and intestine were studied (Thaker et al., 1996). A general dose- and duration-dependent inhibitory trends were found for total ATPase, Na,K-ATPase, Ca-ATPase, Mg-ATPase, Ca/HCO_3-ATPase and Mg^{2+}/HCO_3^--ATPase. However, at the same concentrations initial stimulation of the activity of certain enzymes were also noticed. Chromium might change membrane permeability also and noted effects could be also connected with ROS generation.

Potassium dichromate clearly induced oxidative stress in gills and kidney of European eel *Anguilla anguilla* L. (Ahmad et al., 2006). 1 mM dichromate did not affect catalase and glutathione-S-transferase activities, but increased glutathione peroxides activity and reduced GSH concentrations in gills. Lipid peroxidation, assessed as thiobarbituric acid reactive substances, was intensified in kidney, but no other changes were found in this tissue. DNA integrity, evaluated as DNA strand breaks, was higher in both tissues of dichromate-treated animals (Ahmad et al., 2006).

The activation of free radical processes under fish exposure to chromium compounds may result in change of many physiological parameters. For example, chromium exposure activated lipid peroxidation in tissues of Chinook salmon *Oncorhynchus tshawytscha* and high chromium concentrations significantly impaired fish health (Farag et al., 2006). The kidney was the target organ during chromium exposure – it had gross and microscopic lesions (e.g. necrosis of cells lining kidney tubules) and products of lipid peroxidation were elevated. Noted changes were associated with increased chromium concentrations in the kidney, and reduced fish growth and survival. Authors supposed that chromium was accumulated and induced the lipid peroxidation pathway where fatty acid oxidation and DNA damage (expressed as chromosome changes) occurred and caused cell death and tissue damage (Farag et al., 2006).

Another approach to evaluate free radical induced damage to DNA was used by Knykendall and colleagues (2006). They studied the formation of DNA-protein crosslinks (DPXs) in erythrocytes of largemouth bass *Microproterus salmonoides*, and fathead minnows *Pimephales promelas* exposed to waterborne and dietary hexavalent chromium. Fathead minnow exposure to 2 ppm Cr^{6+} led to significant DPX formation in erythrocytes, with over 140-200% increase above background at 3-4 days, respectively. Largemouth bass exposed similarly were found to have 62% elevation of DPX levels after 4 days. When largemouth bass were fed a diet of minnows injected with 20 mM Cr^{6+} for 5 days, a significant increase of DPXs in erythromycetes was observed. Therefore authors concluded that both waterborne and high dose dietary exposure to Cr^{6+} can lead to DPX formation in erythromycetes of predatory fish species such as bass (Kuykendall et al., 2006).

One of important effects of chromium-containing compounds is connected with modification of immune system which operation involves ROS. Khangarot and colleagues (1999) studied the effects of subtoxic levels of sodium dichromate on humoral and cell-mediated immune responses, blood parameters, succebility to bacterial (*Aeromonas hydrophila*) infection, and macrophage activity in the freshwater air-breathing Indian catfish, *Saccobranchus fassils* Prabakaran and colleagues (2006), investigated the effects of Cr^{6+} exposure on resistance of *Oreochromis mossambicus* to *Aeromonas hydrophila.* They confirmed previous data of Khangarot and colleagues (1999) on decrease of both specific and nonspecific parameters of the immune system which resulted in a lower disease resistance to *Aeromonas hydrophyla*. But these effects were received only for high $K_2Cr_2O_7$ concentrations – 0.5 and 5 mg/l. On the other hand, 0.05 mg/l of Cr^{6+} enhanced disease resistance and both specific and nonspecific immune response to infection. Therefore authors concluded that chromium may modify in concentration-dependent manner immune system demonstrating suppression of stimulatory effects on lymphocytes, lysozyme, phagocytic killing mechanisms, and disease resistance in *Oreochromis mossambicus* (Prabakaran et al., 2006). It can be suggested that dual effects of chromium on immune system exist in other fish as well.

There is a lot of information on the involvement of chromium-containing compounds in toxicity of wastewaters. Because oxidative stress is a common component of most toxicant effects, its parameters are supposed to be useful as biomarkers. In this case, we will refer the only wok clearly demonstrating potential of oxidative stress as a biomarker of environmental pollution by chromium compounds. For example, Parvez and colleagues (2006), investigated parameters oxidative stress in liver, kidney and gills of Indian freshwater fish *Wallago attu* (Bl. and Sch.) during and after fish-kill episode at Panipat, India. They found chromium and copper concentrations could be responsible for fish killing. Both of them are able to enter Haber-Weiss chemistry and therefore the involvement of oxidative stress could be expected. The levels of reduced glutathione and non-protein thiol were higher after the fish-kill episode in all studied tissues. On the other hand, levels of lipid peroxidation products (evaluated as TBARS), and protein carbonyls were lower after mentioned episode, which clearly demonstrated the development of oxidative stress during fish-kill episode. The authors concluded that simultaneous measurement of the physico-chemical parameters of the water samples showed a good correlation with the responses of biomarkers and the environmental chemical stress conditions, and that the markers of oxidative stress may be used to evaluate the environmental toxicity for fish (Parvez et al., 2006).

V.4. Mercury

Mercury exists as cation with an oxidation state +1 (mercurons) and +2 (mercuric). In the environment, mercury may be found in methylmercury form, produced mainly as the result of methylation of inorganic (mercuric) forms by microorganisms in soil and water (Valko et al., 2007). Environmental mercury is ubiquitous and consequently it is practically impossible to avoid exposure to it. Elemental inorganic and organic mercury forms exhibit neurotoxicity, nephrotoxicity, and gastrointestinal toxicity with ulceration and hemorrhage. Mercuric ions have a high affinity to reduced sulfur atoms, particularly to endogenous thiolcontaining molecules, such as glutathione, cysteine, metallothionein, homocysteine, *N*-acetylcysteine, albumin and other proteins. The affinity constant for mercury bonding to thiolate anions is of the order of 10^{15}-10^{20}. For comparison, the affinity constants for mercury bonding to oxygen- (e.g. carbonyl) or nitrogen- (e.g. amino) containing ligands are about 12 orders of magnitude lower. Therefore the biological effects of inorganic or organic mercury are related to their interactions with sulfhydryl-containing residues (Rooney, 2007). Mercuric conjugates of cysteine and glutathione are transportable species at the site of the organic anion transporter.

Because of high affinity of mercury to glutathione, the first can deplete intracellular GSH pool and directly or indirectly causes oxidative stress. It can increase intracellular concentrations of $O_2^{\bullet-}$ and H_2O_2 enhancing ROS-induced damage to cellular components. Mercury compounds disturb calcium metabolism, which may be connected with either their binding to protein thiol groups and/or oxidation of latter. Calcium ions may be released from intracellular stores, particularly from endoplasmic reticulum lumens. The modification of protein-protein contacts between Ca-ATPase molecules of endoplasmic reticulum was hypothesized as a way to increase the calcium release (Lushchak, 1987).

Fish as other animals may accumulate high mercury levels (Salonen et al., 1995; Guallar 2002). High intake of mercury from freshwater by fish and the consequent accumulation of mercury in human body are associated with increased risk of acute myocardial infarction as well as a death resulted from coronary heart disease. Even high content of nonsaturated fatty acids, including docosahexoenoic acid (C 22:6n-3) did not compensate for mercury effects. Therefore, it should be noted that fish oil-derived fatty acids reduce the risk of acute coronary events has a rational basis, but high mercury content in fish could attenuate this protective effect (Valko et al., 2007).

The toxic effects of mercuric ion compounds are well described for higher vertebrates, particularly for mammals and humans, but the information on their effects on fish is extremely limited. As it was mentioned above, concomitant human consumption of mercury with fish oil demonstrates their incorporation into lipid phase and this can result in substantial modification of membrane metabolism. Astroglial cell cultures of both cerebellar and cortical origins demonstrated different sensitivity to methylmercury (Kaur et al., 2007). The latter enhanced ROS production and reduced GSH content. The cerebellar cell cultures were more vulnerable than cortical ones. It was suggested that different mercury toxicity toward different brain regions may be related not only with specific uptake, but also with different content of GSH in cortical and cerebellar cells.

Most fish mercury studies are related with regional brain pathology and behavior changes (Baatrup, 1991; Ribeiro et al., 1995). The biochemical studies are also related to brain operation like inhibition of key enzymes of brain function: acetylcholine esterase (Gill et al., 1990), monoamine oxidase (Ram and Sathyanesan, 1985) and Na,K-ATPase (Carfagna et al., 1995). In experiments with Atlantic salmon *Salmo salar parr* exposed for four months to mercuric chloride methyl mercury accumulated significantly in brain. It did not cause mortality or growth reduction, but resulted in the significant increase of lipid peroxidation products (evaluated as TBARS) and decrease in the activity of SOD and glutathione peroxidases (Berntssen et al., 2003). High mercury concentrations induced pathological damage (vacuolation and necrosis) and decreased monoamine oxidase activity. Comparing with other organs, the brain was particularly susceptible for dietary mercury, while kidney and liver were less sensitive. It should be also noted that low dietary concentrations of mercury induced protective redox defenses in the brain evidenced by the induction of antioxidant enzyme SOD (Berntssen et al., 2003).

V.5. Arsenic

The most common oxidation numbers of arsenic are +5, +3 and -3. It can form both inorganic and organic compounds in the environment and cells. The inorganic arsenics includes arsenide (As^{3+}) and arsenate (As^{5+}). The organic arsenics can be either menthylated (monomethylarsonic acid, MMA), or dimethylated (dimethylasinic acid – DMA) *in vivo* (Valko et al., 2005).

Exposure to arsenic is linked with a risk of developing tumors of the lung, skin, liver, bladder, and kidney (Waalkes et al., 2004). Its concentration in drinking water is usually 0.01-4 mg/l. Arsenate is a phosphate analogue interfering with phosphorylation/dephosphorylation reactions and it also can react with the sulfhydryl groups of proteins and inhibits many biochemical pathways. Now it is widely accepted that arsenic probably does not act as a classical carcinogen, but rather enhances the carcinogenic action of other carcinogens. Arsenic exposure increased the generation of free radicals in cells, which has been linked with the development of arsenic related diseases including cancers. In cells treated with arsenic, ROS and reactive nitrogen species (RNS) are thought to be directly involved in oxidative damage to lipids, proteins, and DNA. In many studies it was found that arsenic-induced generation of free radicals can cause cell damage and death through activation of oxidative sensitive signaling pathways (Shi et al., 2004; Valko et al., 2005).

Several ROS and RNS are known to be involved into generation of dimethylarsinic peroxyl ($(CH_3)_2AsOO^{\circ}$) and dimethylarsinic ($(CH_3)_2As^{\bullet}$) radicals as well as some intermediary arsine reactive species (Rin et al., 1995). It seems that As_2O_3 elevates H_2O_2 production in mitochondria and this may cause damage to cell constituents (Valko et al., 2005).

It is interesting to note that the oxidation of As^{3+} to As^{5+} under physiological conditions results in H_2O_2 formation:

$$H_3AsO_3 + H_2O + O_2 \rightarrow H_3AsO_4 + H_2O_2 \quad (28)$$

Therefore arsenic compounds generating both ROSs and RNSs may be involved in oxidative components particularly lipids, DNA and proteins.

The toxicity studies of arsenic and its different forms have been focused mainly on mammals. For example, arsenite significantly increased both, the concentrations of oxidized glutathione (GSSG) and TBARS in liver of rats, which are clear signatures of oxidative stress development (Kalia et al., 2007). The juvenile mummichogs *Fundulus heteroclitus* whose parents were exposed to 230 ppb arsenic for 10 days immediately prior spawning, the hatchling of exposed fish had 2.8-fold increased incidence of curved or stunned tails (Gonzalez et al., 2006). Arsenic exposure altered in the hatchings the expression of 13 genes. Some of these genes, namely myosin light chain, type II keratin, tropomyosin and parvalbumin, are known to be important during embryogenesis and therefore their differential expression may be linked to the morphological changes observed in the mumichogs hatchlings (Gonzalez et al., 2006). In whole fish and cell cultures arsenic may induce apoptosis (Datta et al., 2007, Wang et al., 2004). In two cell lines, TF (fin cells of *Therapon jarbua*) and TO-2 cells (ovary cells of tilapia) treated with sodium arsenite time- and concentration-dependent cell death was found with line specificity. The DNA-fragmentation analysis and the flow cytometric analysis of cell cycle progression demonstrated clearly that most cells were killed via apoptosis. Since antioxidants, *N*-acetyl-cysteine and dithiothreitol significantly prevented apoptosis in TF cells, it was concluded that ROS are involved in arsenite-induced apoptotic cell death (Wang et al., 2004). More detail investigation of arsenic effects on whole Indian catfish *Clarias batrachus* was carried out by Datta and colleagues (2007). They found that subletal arsenic exposure (0.50 μM) altered serum aspartate aminotransferase and alkaline phosphatase activities. It also reduced hepatocyte protein content and suppressed proliferation of hepatocytes. Histological studies demonstrated arsenic-induced changes characterized by dilated sinusoids, formation of intracellular edema, megalocytosis, vacuolation and appearance of hepatic cells with distorted nuclei. Electron microscopy showed hyperplasia and hypertrophy of mitochondria, development of dilated rough endoplasmic reticulum and changes in peroxisome size with duration of arsenic exposure. Degeneration of mitochondrial crystal, condensation of chromatin, staining with annexin V, and, finally a specific DNA ladder characteristic taken together let authors to conclude on the development of apoptosis induced by fish exposure to arsenite. That was also confirmed by the using specific inhibitors of caspases, particularly it seems caspase 3 pathway was involved in arsenite-induced apoptosis of hepatocytes (Datta et al., 2007).

Schlenk and colleagues (1997) studied the effects of arsenite, arsenate and the herbicide monosodium methyl arsonate on hepatic metalloprotein expression and lipid peroxidation in channel catfish. They found dose-dependent increase in metallothionein levels, while hepatic lipid peroxidation (determined as TBARS) and glutathione concentrations were unaltered by any of the arsenical treatments. It demonstrates the lack of correlation between metallothionein expression and oxidative stress.

We also investigated the effects of arsenite treatment on oxidative stress markers and antioxidant defense in goldfish liver (Bagnyukova et al., 2007). The treatment during 1-4 days did not affect concentrations of TBARS and protein carbonyls (measures of lipid and protein ROS- induced oxidation, respectively). However, it increased the oxidation of glutathione, total glutathione concentration and lipid peroxides clearly showing the development of oxidative stress. The activities of the main antioxidant enzymes – superoxide dismutase, catalase and glutathione peroxidase were also enhanced demonstrating an antioxidant response. These results clearly show that arsenite casued oxidative stress in goldfish liver.

Taking into account the described above effects of arsenic compounds on fish it is easy to understand, that this element is able to enter cellular metabolism and in some cases it enhances the production of free radicals. The latter can modify virtually all cellular constituents, including membranes, leading to cell death via necrosis or apoptosis and the discrimination between these both ways depends on many circumstances.

V.6. Other Transition Metals

Several other ions of transition metals not mentioned above may have certain biological significance and/or be toxic. They are cobalt, vanadium and nickel. These elements are found in individual organisms and food chains, but the information on the levels and effects on fish is extremely scarce. For example, Soares and colleagues (2007) found that vanadium affected the concentrations of lipid peroxide products and activities of certain antioxidant enzymes in fish *Sparus aurata* heart. Vanadates were accumulated into mitochondria, inhibited mitochondrial oxygen consumption, induced mitochondrial depolarization, and changed the redox state of complex III. Moreover, the efficiency depended very much on used species of vanadate – oligomeric vanadium species were much more effective than monomeric. Authors concluded that mitochondria is a toxicological target for vanadate, but resultative effect depends on vanadate oligomeric state (Soares et al., 2007). Clearly, the investigation of effects of these metals is in need.

VI. Conclusions and Perspectives

The broad distribution of transition metals in biosphere and their redistribution due to human activity have called serious attention to them. It is long known that iron, copper and chromium are essential elements involved in many cellular processes like the first two are the components of electron transport chain. In fact, there is no chance to avoid biosphere pollution by transition metals. One may expect at least for the decrease of this pollution, but the question on acceptable limits is still opened. In addition, we may look for ways to reduce their negative effects on individual organisms and/or food chains. It became more and more clear, that low concentrations of analyzed above elements and their derivatives are stimulating. In very simple experiments using water flea *Daphnia magna* in our students' laboratory we have found that low concentrations of compounds of iron, chromium and copper stimulated growth and increased the fecundity of this crustacean (not published). Similar simulating effects were found in studying of growth of mycelium of edible mushrooms Oyster mushroom *Pleurotus ostreatus* and shiitake *Lentinus edodes* (not published). There is also the information that even elements with unknown biological function like arsenic and mercury up to some concentrations may improve the organism's state. Therefore the next relationship between concentrations of transition element ions and transition of organism may be given (Figure 4): these elements at low concentrations stimulate organism because many of the ions are essential elements. Even arsenic and mercury may affect positively via mild stress stimulating energy provision and antioxidant potential. There is clear homology with gerontological conception „hormesis", when mild stress extends lifespan of organisms. Now the stimulating effects of stress induced by any method became understable because it „trains" organism to cope with stronger stresses. The increase of efficiency of energy provision systems and coping with free radicals may be among main mechanisms responsible for these effects. The second part of the graph at Figure 4 is optimal concentration range. In this case, all systems which need essential elements are saturated and organism operates under optimal conditions. That is "normal state" when all functions are realized in the optimal way. The increase of concentrations above optimal often does not cause negative effects because there are specific and nonspecific systems neutralizing excess of metal ions. This is buffered zone. The tolerated interval depends very much not only from the ion type, but also on transport and defense systems. Evolutionary conservative metallothioneins and chaperons are probably the best studied systems responsible for the increase of organism tolerance to metal ions. They have high affinity to transition metal ions and usually are induced by them. These proteins also have relatively low specificity to metal ions and therefore may often bind several their types. But not only binding of ions may be the protective mechanism. Because these ions stimulate ROS production, they may also stimulate antioxidant systems. It is well established that these ions enhance for example, glutathione synthesis, activities of antioxidant enzymes, etc. The higher concentrations of transient metal ions became toxic and one of the mechanisms of their toxicity was analyzed above. It is now more or less clear that oxidative stress usually is an element of transition metal ion toxicity. However, it is impossible to avoid biosphere pollution by metal ions. Therefore, the questions are: how to reduce their negative effects on organisms, how to increase tolerated and even more – optimal concentrations? To respond these questions we have to investigate the mechanisms of their deleterious effects.

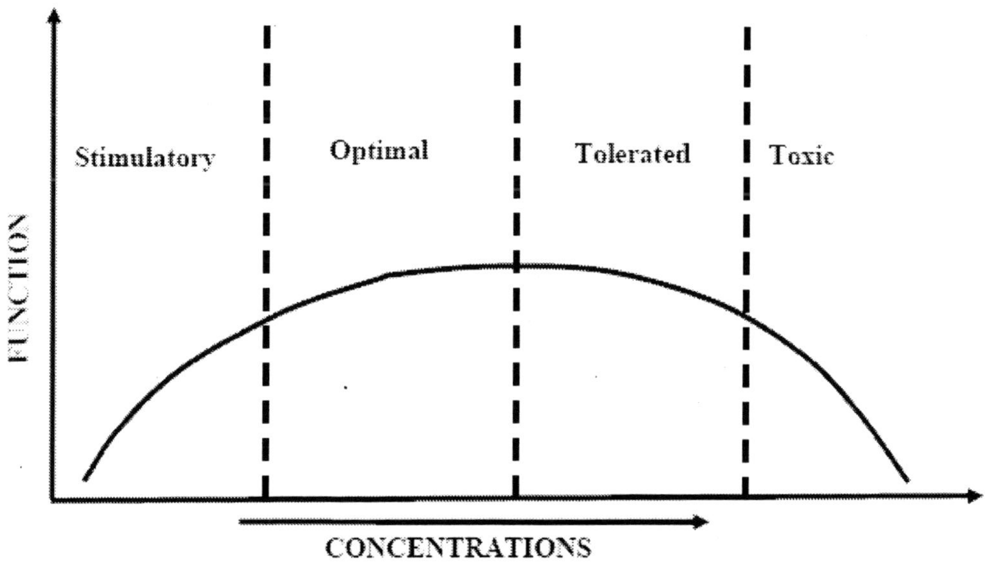

Figure 4. Interaction between concentrations of transition metal ions and integral indexes of organism function.

That is the first step. Basing on fundamental knowledge we will be able to identify and further investigate the possible targets for this sort of prevention of deleterious effects of metal ions. For this we can use well known fact that pre-adaptation to low doses of toxicants or mild stress which increase organism tolerance to higher toxicant concentrations or more intensive stress.

So-called cross-adaptations are other way to reduce negative effects of metal ions, but it is not well studied in animals. However, the issue has been extensively studied in microorganisms. If some group of toxicants or other negative effectors like temperature, aging, toxins share similar mechanisms of negative effects, the treatment with one of stressors can increase tolerance to another (Lushchak, 2001). Because the induction of oxidative stress by transition metal ions is one of the mechanisms involved in their toxicity, the increase of antioxidant potential can be beneficial for organisms. One of the possible mechanisms may be connected with an application of natural and synthetic antioxidants, their precursors and functional analogs. One may expect that even physical stress like exercise may enhance organism tolerance to transition metal ions, but it needs further investigation.

The tissue specificity of metal ion distribution cannot be ignored also. For example, it is well known that copper is selectively accumulated in liver, while mercury does in brain. Therefore, the first may cause liver cirrhosis and the second – neurological disorders. But all systems in organism are interrelated and some tissues may be not sensitive to high concentrations, but other can not resist even very low ones.

One more important issue, which it critically important for understanding of transient metal toxicity is their migration in food chains. Transient metal ions cannot be destroyed by any organism and therefore can be either used for certain purposes or kept in secure form. They may be bound in cytoplasm to relatively nonspecific metallothioneins and chaperons or more specific proteins like ferritin for iron and ceruloplasmin for copper. A sequestration is other way to keep them, for example, into vacuole or other cellular organelles. The binding and sequestration are very effective mechanisms to control metal ion intracellular concentrations. For example, as it was mentioned above, under normal conditions less than one free copper ion may be found in the single cell. The cellular reduction and oxidation may be also responsible for the fine control of free ions at loosely low concentrations and to avoid their negative effects. Therefore it can be concluded that final effect of transition metal ions on organism involves their transportation, binding, sequestration and metabolism and that organisms possess powerful and multilevel systems to cope with them. The efficiency of defense systems may be enhanced and the study of mechanisms responsible for toxicity is the avenue for future.

ACKNOWLEDGEMENT

The authors would like to thank B. Rovenko for the technical assistance with manuscript preparation.

CITED LITERATURE

Ahmad, I., Maria, V.L., Oliveira, M., Pacheco, M., Santos, M.A., 2006. Oxidative stress and genotoxic effects in gill and kidney of *Anguilla anguilla* L. exposed to chromium with or without pre-exposure to β-naphthoflavone. Mut. Res.,608(1): 16-28.

Aisen, P., Enns, C., Wessling-Resnick, M., 2001. Chemistry and biology of eukaryotic iron metabolism. *Int. J. Biochem. Cell. Biol.* 33(10): 940-959.

Baatrup, E., 1991. Structural and functional effects of heavy metals on the nervous system, including sense organs, of fish. *Comp. Biochem. Physiol.* C100: 253-257.

Banci, L., Bertini, I., Durazo, A., Girotto, S., Gralla, E.B., Martinelli, M., Valentine, J.S., Vieru, M., Whitelegge, J.P., 2007. Metal-free superoxide dismutase forms soluble oligomers under physiological conditions: A possible general mechanism for familial *ALS. Proc. Natl. Acad. Sci.* U S A. 104(27):11263-11267.

Bagchi, D., Stohs, S. J., Downs, B. W., Bagchi, M., Preuss, H. G. 2002. Cytotoxicity and oxidative mechanisms of different forms of chromium. *Toxicology.*180(1): 5-22.

Bagnyukova, T.V., Chahrak, O.I., Lushchak, V.I., 2006. Coordinated response of goldfish antioxidant defenses to environmental stress. *Aquat. Toxicol.* 78: 325-331.

Bagnyukova, T.V., Luzhna, L.I., Pogribny, I.V, Lushchak, V.I. 2007. Oxidative stress and antioxidant defenses in goldfish liver in response to short-term exposure to arsenite. *Environ. Mol. Mutagen.* 48(8): 658-665.

Baker, R. T. M., Handy, R. D., Davies, S. J., Snook, J. C. 1998. Chronic dietary exposure to copper affects growth, tissue lipid peroxidation, and metal composition of the grey mullet, *Chelon labrosus*. Mar. Environ. Res. 45(4-5): 357-365.

Baker, R. T. M., Martin, P., Davies, S. J., 1997. Ingestion of sub-lethal levels of iron sulphate by African catfish affects growth and tissue lipid peroxidation. *Aquat. Toxicol.* 40(1): 51-61.

Berntssen, M.H.G., Aatland, A., Handy, R.D., 2003. Chronic dietary mercury exposure causes oxidative stress, brain lesions, and altered behaviour in Atlantic salmon (*Salmo salar*) parr. *Aquat. Toxicol.* 65(1): 55-72.

Blanchard, J., Grosell, M. 2006. Copper toxicity across salinities from freshwater to seawater in the euryhaline fish *Fundulus heteroclitus*: Is copper an ionoregulatory toxicant in high salinities? *Aquat. Toxicol.* 80(2): 131-139.

Blasiak, J., Kowalik, J., 2000. A comparison of the *in vitro* genotoxicity of tri- and hexavalent chromium. *Mut. Res.* 469(1): 135-145.

Bosgelmez, I.I., Guvendik, G., 2004. Effects of taurine on oxidative stress parameters and chromium levels altered by acute hexavalent chromium exposure in mice kidney tissue. *Biol Trace Elem Res.* 102(1-3): 209-225.

Bury, N. R., Grosell, M., Wood, C. M., Hogstrand, C., Wilson, R. W., Rankin, J. C., Busk, M., Lecklin, T., Jensen, F. B. 2001. Intestinal iron uptake in the European flounder (*Platichthys flesus*). *J. Exp. Biol.* 204: 3779-3787.

Bury, N.R., Walker, P.A., Glover, C.N., 2003. Nutritive metal uptake in teleost fish. *J. Exp. Biol.* 206(Pt 1): 11-23.

Carfagna, M. A., Ponsler, G. D., Muhoberac, B. B., 1996. Inhibition of ATPase activity in rat synaptic plasma membranes by simultaneous exposure to metals. *Chem.-Biol. Interac.* 100(1): 53-65.

Carriquiriborde, P., Handy, R. D., Davies, S. J., 2004. Physiological modulation of iron metabolism in rainbow trout (*Oncorhynchus mykiss*) fed low and high iron diets. *J. Exp. Biol.* 207: 75-86.

Coogan, T.P., Motz, J., Snyder, C.A., Squibb, K.S., Costa, M., 1991. Differential DNA-protein crosslinking in lymphocytes and liver following chronic drinking water exposure of rats to potassium chromate. *Toxicol. Appl. Pharmacol*. 109(1): 60-72.

Datta, S., Saha, D. R., Ghosh, D., Majumdar, T., Bhattacharya, S., Mazumder, S. 2007. Sub-lethal concentration of arsenic interferes with the proliferation of hepatocytes and induces *in vivo* apoptosis in *Clarias batrachus* L. *Comp. Biochem. Physiol.* 145(3): 339-349.

De Boeck, G., Van der Ven, K., Meeus, W., Blust, R., 2007. Sublethal copper exposure induces respiratory stress in common and gibel carp but not in rainbow trout. *Comp. Biochem. Physiol.,* C144(4): 380-390.

Demple, B., 1990. Oxidative DNA damage: repair and inducible cellular responses. *Prog. Clin. Biol. Res.* 340A: 155-167.

Demple, B., Sung, J.S., 2005. Molecular and biological roles of Ape1 protein in mammalian base excision repair. DNA Repair (Amst). 4(12): 1442-1449.

Dunlop, R.A., Rodgers, K.J., Dean, R.T., 2002. Recent developments in the intracellular degradation of oxidized proteins. Free Radic. *Biol. Med.* 33(7): 894-906.

Farag, A. M., May, T., Marty, G. D., Easton, M., Harper, D. D., Little, E. E., Cleveland L., 2006. The effect of chronic chromium exposure on the health of Chinook salmon (*Oncorhynchus tshawytscha*). *Aquat. Toxicol.* 76(3-4): 246-257.

Fernandes, M.A., Santos, M.S., Alpoim, M.C., Madeira, V.M., Vicente, J.A., 2002. Chromium(VI) interaction with plant and animal mitochondrial bioenergetics: a comparative study. *J. Biochem. Mol. Toxicol.* 16(2): 53-63.

Gill, T., Tewari, H., Pandee, J., 1990. Use of the fish enzyme system in monitoring water quality: effects of mercury on tissue enzymes. *Comp. Biochem. Physiol.* C97: 287-292.

Gonzalez, H. O., Roling, J. A., Baldwin, W. S., Bain, L. J., 2006. Physiological changes and differential gene expression in mummichogs (*Fundulus heteroclitus*) exposed to arsenic. *Aquat. Toxicol.*,77(1): 43-52.

Gospodaryov, D.V., Bailyak, M.M., Lushchak, V.I., 2005. Free radical *in vitro* inactivation of glucose-6-phosphate dehydrogenase from the yeast *Saccharomyces cerevisiae. Ukrainian Biochem. J.*, 77(1): 54-60.

Gospodaryov, D.V., Lushchak, V.I., 2004. The influence of iron on the antioxidant enzyme activities in yeast *Saccharomyces cerevisiae*, Ukrainian Biochem. J., 76(6): 121-126.

Gospodarev, D.V., Lushchak, V.I., 2005. Iron metabolism in yeasts, *Ukrainian Biochem. J.*, 77(3): 5-19.

Guallar, E., Sanz-Gallardo, M.I., van't Veer, P., Bode, P., Aro, A., Gomez-Aracena, J., Kark, J.D., Riemersma, R.A., Martin-Moreno, J.M., Kok, F.J. 2002. Heavy metals and myocardial infarction study group. Mercury, fish oils, and the risk of myocardial infarction. *N. Engl. J. Med.* 347(22): 1747-1754.

Gunaratnam, M., Grant, M. H., 2001. The role of glutathione reductase in the cytotoxicity of chromium (VI) in isolated rat hepatocytes. *Chem.-Biol. Interact.* 134(2): 191-202.

Halliwell, B., Gutteridge, J.M.C., 1989. *Free Radicals in Biology and Medicine* (2nd ed.). Oxford, UK: Clarendon Press.

Hoyle, I., Shaw, B.J., Handy, R.D., 2007. Dietary copper exposure in the African walking catfish, *Clarias gariepinus*: Transient osmoregulatory disturbances and oxidative stress. *Aquat. Toxicol.* 83(1): 62-72.

Kalia, K., Narulaa, G. D., Kannan, G.M., Flora, S.J.S., 2007. Effects of combined administration of captopril and DMSA on arsenite induced oxidative stress and blood and tissue arsenic concentration in rats. *Comp. Biochem. Physiol.* C144: 372–379.

Kamunde, C., Clayton, C., Wood, C. M. 2002. Waterborne vs. dietary copper uptake in rainbow trout and the effects of previous waterborne copper exposure. *Am. J. Physiol.* 283: R69-R78.

Kapiotis, S., Hermann, M., Exner, M., Laggner, H., Gmeiner, B.M., 2005. Copper- and magnesium protoporphyrin complexes inhibit oxidative modification of LDL induced by hemin, transition metal ions and tyrosyl radicals. *Free Radic. Res.* 39(11): 1193-1202.

Kaplan, J., McVey Ward, D., Crisp, R.J., Philpott, C.C., 2006. Iron-dependent metabolic remodeling in *S. cerevisiae. Biochim. Biophys. Acta.* 1763(7): 646-651.

Kaur, P., Aschner, M., Syversen, T., 2007. Role of glutathione in determining the differential sensitivity between the cortical and cerebellar regions towards mercury-induced oxidative stress. *Toxicology.* 230(2-3): 164-177.

Khangarot, B.S., Rathore, R.S., Tripathi, D.M., 1999. Effects of chromium on humoral and cell-mediated immune responses and host resistance to disease in a freshwater catfish, *Saccobranchus fossilis* (Bloch). Ecotoxicol. Environ. Saf. 43(1): 11-20.

Kuykendall, J. R., Miller, K. L., Mellinger, K. N., Cain, A. V., 2006. Waterborne and dietary hexavalent chromium exposure causes DNA-protein crosslink (DPX) formation in erythrocytes of largemouth bass (*Micropterus salmoides*). *Aquat. Toxicol.* 78(1): 27-31.

Lanno, R. P., Slinger, S. J., Hilton, J. W., 1985. Maximum tolerable and toxicity levels of dietary copper in rainbow trout (*Salmo gairdneri* Richardson). *Aquaculture.* 49(3-4): 257-268.

Leonard, S. S., Harris, G. K., Shi, X., 2004. Metal-induced oxidative stress and signal transduction. *Free Radic. Biol. Med* 37(12): 1921-1942.

Liu, H., Wang, W., Zhang, J.-F., Wang X-R., 2006. Effects of copper and its ethylenediaminetetraacetate complex on the antioxidant defenses of the goldfish, *Carassius auratus. Ecotoxicol. Environ. Safety.* 65(3): 350-354.

Lloyd, D.R., Phillips, D.H., Carmichael, P.L., 1997. Generation of putative intrastrand cross-links and strand breaks in DNA by transition metal ion-mediated oxygen radical attack, *Chem. Res. Toxicol*. 10: 393–400.

Lushchak, V.I., 1987. The possible physiological role of the oligomeric organization of Ca-ATPase of sarcoplasmic reticulum. *Biol. Sciences* (USSR). 10: 100-109.

Lushchak, V.I., 2001. Oxidative stress and mechanisms of protection against it in bacteria. *Biochemistry* (Moscow) 66: 476-489.

Lushchak, V.I., 2002. Oxidative stress in bacteria. In: Oxidative stress at molecular, cellular and organ levels. Johnson P., Boldyrev A., *eds. Research Signpost:* 45-65.

Lushchak, V.I., 2007. Free radical oxidation of proteins and its relationship with functional state of organisms. *Biochemistry* (Moscow). 72 (8): 995-1017.

Mehta, R., Templeton, D. M., O'Brien, P. J., 2006. Mitochondrial involvement in genetically determined transition metal toxicity: II. Copper toxicity. *Chem.-Biol. Interact.* 163(1-2): 77-85.

Parvez, S., Pandey, S., Ali, M., Raisuddin, S. 2006. Biomarkers of oxidative stress in *Wallago attu* (Bl. and Sch.) during and after a fish-kill episode at Panipat, India. *Sci. Total Environ.* 368(2-3): 627-636.

Pedrajas, J. R., Peinado, J., López-Barea, J., 1995. Oxidative stress in fish exposed to model xenobiotics. Oxidatively modified forms of Cu,Zn-superoxide dismutase as potential biomarkers. *Chem.-Biol. Interact.* 98(3): 267-282.

Perez-Benito, J. F., 2006. Effects of chromium(VI) and vanadium(V) on the lifespan of fish. *J. Trace Elem. Med. Biol.* 20(3): 161-170.

Ponka, P., 1999. Cellular iron metabolism. *Kidney Int. Suppl.* 69: S2-S11.

Pourahmad, J., O'Brien, P.J., 2001. Biological reactive intermediates that mediate chromium (VI) toxicity. *Adv. Exp. Med. Biol.* 500: 203-207.

Pourahmad, J., O'Brien, P.J., Jokar, F., Daraei, B., 2003. Carcinogenic metal induced sites of reactive oxygen species formation in hepatocytes. Toxicol. *In Vitro*. 17(5-6): 803-810.

Prabakaran, M., Binuramesh, C., Steinhagen, D., Michael, R.D., 2006. Immune response and disease resistance of *Oreochromis mossambicus* to *Aeromonas hydrophila* after exposure to hexavalent chromium. Dis. Aquat. Organ. 68(3): 189-96.

Rae, T.D., Schmidt, P.J., Pufahl, R.A., Culotta, V.C., O'Halloran, T.V., 1999. Undetectable intracellular free copper: the requirement of a copper chaperone for superoxide dismutase.*Science.* 284(5415): 805-808.

Ram, R., Sathyanesan, A., 1985. Mercurial induced brain monoamine oxidase inhibition in the teleost *Channa punctarus* (Bioch). Bull. Environ. Contam. Toxicol. 35, 620-626.

Requena, J.R., Levine, R.L., Stadtman, E.R., 2003. Recent advances in the analysis of oxidized proteins. *Amino Acids.* 25(3-4): 221-226.

Ribeiro, C., Fernandes, L., Carvlaho, C., Cardoso, R., Tucatti, N., 1995. Acute effects of mercuric chloride on the olfactory epithelium of *Trichomycterus brasiliensis. Ecotoxicol. Environ. Saf.* 31: 104-109.

Rin, K., Kawaguchi, K., Yamanaka, K., Tezuka, M., Oku, N., Okada, S., 1995. DNA-strand breaks induced by dimethylarsinic acid, a metabolite of inorganic arsenics, are strongly enhanced by superoxide anion radicals. *Biol. Pharm. Bull.* 18(1): 45-48.

Rooney, J. P.K., 2007. The role of thiols, dithiols, nutritional factors and interacting ligands in the toxicology of mercury. *Toxicology.* 234(3): 145-156.

Rotilio, J., Stella, A.M., 2006. Molecular basis of neurodegenerative diseases. *Ital. J. Biochem.* 55(3-4): 189-193.

Salonen, J.T., Seppanen, K., Nyyssonen, K., Korpela, H., Kauhanen, J., Kantola, M., Tuomilehto, J., Esterbauer, H., Tatzber, F., Salonen, R., 1995. Intake of mercury from fish, lipid peroxidation, and the risk of myocardial infarction and coronary, cardiovascular, and any death in eastern Finnish men. *Circulation.* 91(3): 645-655.

Sheehan, J. P., Swerdlow, R. H., Miller, S. W., Davis, R. E., Parks, J. K., Parker, W. D., Tuttle, J. B., 1997. Calcium homeostasis and reactive oxygen species production in cells transformed by mitochondria from individuals with sporadic Alzheimer's disease, *J. Neuroscience.* 17(12): 4612-4622

Senthilnathan, P., Padmavathi, R., Magesh, V., Sakthisekaran, D., 2006. Chemotherapeutic efficacy of paclitaxel in combination with *Withania somnifera* on benzo(a)pyrene-induced experimental lung cancer. *Cancer Sci.* 97(7): 658-664.

Schlenk, D., Wolford, L., Chelius, M., Steevens, J., Chan, K. M., 1997. Effect of arsenite, arsenate, and the herbicide monosodium methyl arsonate (MSMA) on hepatic metallothionein expression and lipid peroxidation in channel catfish. *Comp. Biochem. Physiol.* C118(2): 177-183.

Shcherbatykh, I., Carpenter, D.O., 2007. The role of metals in the etiology of Alzheimer's disease. *J. Alzheimers Dis.* 11(2): 191-205.

Shi, X., Dalal, N. S., 1990. Evidence for a fenton-type mechanism for the generation of $^{\cdot}OH$ radicals in the reduction of Cr(VI) in cellular media. *Arch. Biochem. Biophys.* 281(1): 90-95.

Shi, H., Shi, X., Liu, K.J. 2004. Oxidative mechanism of arsenic toxicity and carcinogenesis. *Mol. Cell. Biochem.* 255(1-2): 67-78.

Sies, H., 1985. Oxidative Stress. Academic Press, London.

Sies, H., Strategies of antioxidant defense. 1993. *Eur J Biochem.* 215(2): 213-219.

Soares, S. S., Gutiérrez-Merino, C., Aureliano, M. 2007. Mitochondria as a target for decavanadate toxicity in *Sparus aurata* heart Aquat. Toxicol., 83(1) 2007: 1-9.

Stadtman, E.R., Levine, R.L., 2000. Protein oxidation. Ann. NY Acad. Sci. 899: 191-208.

Subashini, P., Manavalaramanujam, R., Ramesh, M., Geetha, N., 2005. Changes in selected biomarkers in freshwater teleost fish, *Cyprinus carpio* var. communis exposed to sublethal concentrations of chromium sulphate toxicity. *J. Environ. Sci. Eng.* 47(1): 65-68.

Susa, N., Ueno, S., Furukawa, Y., Sugiyama, M. 1996. Protective effect of vitamin E on chromium (VI)-induced cytotoxicity and lipid peroxidation in primary cultures of rat hepatocytes. *Arch. Toxicol.* 71(1-2): 20-24.

Susa, N., Ueno, S., Furukawa, Y., Sugiyama, M., 1997. Protective effect of deferoxamine on chromium (VI)-induced DNA single-strand breaks, cytotoxicity, and lipid peroxidation in primary cultures of rat hepatocytes. *Arch. Toxicol.* 71(6): 345-350.

Tagliari, K. C., Vargas, V. M. F., Zimiani, K., Cecchini, R., 2004. Oxidative stress damage in the liver of fish and rats receiving an intraperitoneal injection of hexavalent chromium as evaluated by chemiluminescence. Env. *Toxicol. Pharmacol.* 17(3): 149-157.

Thaker, J., Chhaya, J., Nuzhat, S., Mittal, R., Mansuri, A. P., Kundu, R., 1996. Effects of chromium(VI) on some ion-dependent ATPases in gills, kidney and intestine of a coastal teleost *Periophthalmus dipes. Toxicology.* 112(3): 237-244.

Ueno, S., Sugiyama, M., Susa, N., Furukawa, Y., 1995. Effect of dimethylthiourea on chromium (VI)-induced DNA single-strand breaks in Chinese hamster V-79 cells. *Mutat. Res.* 346(4): 247-253.

Valko, M., Leibfritz, D., Moncol, J., Cronin, M.T., Mazur, M., Telser, J., 2007. Free radicals and antioxidants in normal physiological functions and human disease. *Int. J. Biochem. Cell. Biol.* 39(1): 44-84.

Van Ho, A., Ward, D.M., Kaplan, J., 2002. Transition metal transport in yeast. *Annu. Rev. Microbiol.* 56: 237-261.

Vladimirov, Y.A., 2002. Free radicals in cell life – two sides of the same coin. In: Oxidative stress at molecular, cellular and organ levels. Johnson P., Boldyrev A., eds. *Research Signpost:* 13-44.

Waalkes, M.P., Liu, J., Ward, J. M., Diwan, B.A., 2004. Mechanisms underlying arsenic carcinogenesis: hypersensitivity of mice exposed to inorganic arsenic during gestation Toxicology. 198(1-3): 31-38.

Walker, R.L., Fromm, P.O., 1976. Metabolism of iron by normal and iron deficient rainbow trout. *Comp. Biochem. Physiol.* 55(4): 311-318.

Wang, Y.-C., Chaung R.-H., Tung, L.-C. 2004. Comparison of the cytotoxicity induced by different exposure to sodium arsenite in two fish cell lines. Aquat. Toxicol. 69(1): 67-79.

Watanabe, T., Kiron, V., Satoh, S., 1997. Trace minerals in fish nutrition. *Aquaculture.* 151(1-4): 185-207.

In: Aquatic Toxicology Research Focus
Editor: Elias P. Svensson, pp. 31-49
ISBN: 978-1-60456-192-0

Chapter 2

PHARMACEUTICALS IN WATER AND THEIR POTENTIAL IMPACT ON REUSE

F. Pomati [1][1]***, B.P. Burns*** [1]***, E. Zuccato*** [2]***, B.A. Neilan*** [1]***, and S.J. Khan*** [3]

[1]School of Biotechnology and Biomolecular Sciences, The University of New South Wales, 2052 Australia,

[2]Department of Environmental Health Sciences, Mario Negri Institute for Pharmacological Research, 20157 Milan, Italy,

[3]Centre for Water and Waste Technology, The University of New South Wales, 2052 Australia

ABSTRACT

Recent studies have documented the presence of a wide variety of pharmaceuticals in the environment. Active pharmaceutical ingredients (APIs) can enter water systems via several routes, but municipal sewage effluent is considered to be one of the most important. Pharmaceuticals have been detected both in sewage influent and effluent, in surface water and groundwater and increasingly also in drinking water. Information on the long-term effects of the presence of APIs on aquatic life or human health is largely lacking. The aim of this chapter is to review the ecotoxicological work dealing with environmental contamination by pharmaceuticals, with particular regard to the fate and effects of APIs in urban water cycles, and to provide future perspectives and research needs. Pharmaceuticals are widespread contaminants, entering the environment from a myriad of scattered points. Patients, in the case of drugs for human use, or animals for veterinary drugs, are the main sources of contamination. The first step to be taken for a thorough environmental and human health risk assessment is to develop standardised methods for toxicological evaluation, as single compound and combined in mixtures of APIs that are representative for the real environment. *In vitro* and *in vivo* studies suggest that APIs, taken singularly or in combinations, and in concentrations close to those detected in the environment, may have ecotoxicological effects. Preliminary studies indicate that mixtures of APIs may also pose a risk for human health. Once thorough

---

[1] Author for correspondence: f.pomati@unsw.edu.au.

toxicological and ecotoxicological effects have been thoroughly investigated, APIs of particular environmental and toxicological concern can be identified. Sewage treatment processes are an important point in the control of contamination, but conventional treatment plants are generally not able to efficiently abate a substantial number of waterborne APIs. Several variables play a role, however, in the processes of sewage treatment, and could be specifically adjusted to improve the efficiency of API abatement, mitigating the potential environmental and human exposure.

Overview

"The increasing worldwide contamination of freshwater systems with thousands of industrial and natural chemical compounds is one of the key environmental problems facing humanity" [Schwarzenbach et al. 2006]. Clean water is an enduring symbol of the force of life and a basic right to human needs.

Pollution is changing along with the modernisation of human life. The demand for active pharmaceutical ingredients (APIs) has increased tremendously on a global scale during the last two decades, and it is expected to rise further in the future. Presently, therapeutic drugs are sold worldwide via a market with an annual production of tens of thousands of tons that covers both human and veterinary practices. The pharmaceutical industry is attaining more effective APIs by designing for increased potency, bioavailability and degradation resistance. As a result of excretion, discharge, or disposal, almost 100 pharmaceutically active compounds from various therapeutic classes have been detected up to μg/L levels in sewage, surface, ground and drinking water worldwide [Zuccato et al. 2000, Khan and Ongerth 2002, Ternes and Joss 2006]. These comprise antibiotics, anaesthetics, anti-inflammatories, anti-tumour compounds, steroidal hormones, lipid reducing agents, anti-diuretics, anti-depressives, analgesics, antibiotics and x-ray contrast media, among others. Furthermore, a range of illicit drugs such as cocaine, amphetamines, and THC, has also been reported [Zuccato et al. 2005b]. Such documented evidence raises novel and intricate toxicological and environmental issues.

The mechanisms of action of most active pharmaceutical agents are well known and described. The ultimate effects of pharmaceutical mixtures and their metabolites in the environment are, however, unknown. Environmental concentrations of APIs are remarkably smaller than the corresponding therapeutic doses employed for medical treatments, and acute toxicity tests have often failed to detect the subtle action elicited by drugs at low dosage. Nevertheless, exposure to these chemicals in the environment presents peculiar and potentially hazardous characteristics. APIs are ordinarily detected in complex mixtures of unrelated molecules, and interactions between classes of drugs represent a serious toxicological issue, still also largely unresolved in medicine. In addition, drugs may affect chronically a variety of organisms in different stages of development and for which no pharmacological action is known. Concerns include the role of some drugs to act as endocrine disruptors and the capability of complex drug mixtures of inducing genetic alteration/mutation [Anway et al. 2005], or the possible danger posed by waterborne APIs in water recycling for human and agricultural use [Ternes and Joss 2006].

The addition to water of a cocktail of trace quantities of pharmaceuticals, compounds that are designed to exhibit potent physiological activity, is an important and emerging issue. We now know that certain pharmaceuticals can persist in the environment and, via drinking water, may be subject to unintended human exposure. Particular concern has been expressed regarding unintended exposure of pregnant mothers and children. Some of these agents are beginning to be associated with adverse developmental effects in aquatic organisms at environmentally relevant concentrations that we normally consider to be insignificant and harmless. Furthermore, we know almost nothing about the impacts of long-term human exposure to low-dose mixtures of pharmaceuticals or of low-dose pharmaceuticals mixed with other low-dose synthetic pollutants, but that the little we do know gives reason for concern. Scientific questions mainly address negative impacts on aquatic species, but naturally other concerns arise about human health. Among the most significant are those relating to the potential for developmental impairment by trace quantities of pharmaceuticals in drinking water. Based on our previous work, and inspired by a recent review that we suggest as a further reading [Khetan and Collins 2007], we summarise this subject area in the present chapter. We have chosen not to discuss hormones and other endocrine disrupting chemicals in the present review, as a large body of literature is currently available.

Origin and Causes of Contamination

Pharmaceuticals are widespread contaminants, entering the environment from a myriad of scattered points. Patients, in the case of drugs for human use, or animals for veterinary drugs, are the main sources of contamination. Pharmaceuticals can be excreted as the parent compound and/or metabolites in urine and faeces. Several pharmaceuticals widely used in human medicine are excreted unchanged or as active metabolites in high percentages, and continuously discharged into domestic wastewaters [Kummerer 2004a]. Unwanted or expired medications are often disposed of, directly in wastewater, and this is recommended practice in many countries. Accordingly, several pharmaceuticals are able to reach sewage treatment plants (STPs) in substantial amounts and, if they escape degradation, can be released into surface waters. Most wastewater biological treatment processes do not completely remove therapeutic compounds [Ternes 1998, Stumpf et al. 1999,Khan and Ongerth 2005]. Recent analytical studies in Italy confirm that several pharmaceuticals are poorly removed in STPs [Castiglioni et al. 2006]. Moreover, some pharmaceuticals can be excreted as conjugates, and may be transformed back into the active parent compound by cleavage during treatment in STPs [Heberer 2002]. In the case of drugs for human use, STPs are therefore the most important point source of contamination. In addition to improper disposal of expired medications, manufacturing facilities are other important local point sources [Zuccato et al. 2000]. Aquaculture industries and run-off from farms play an additional significant role in the presence of some specific drugs in the environment [Calamari et al. 2003].

The use of illicit drugs such as cocaine, heroin, amphetamine-like stimulants, and marijuana, may be similarly excreted and flow into and through wastewater treatment plants [Zuccato et al. 2005b, Castiglioni et al. 2006]. Since some drugs may have long half-lives in the environment, they can accumulate, reaching detectable and biologically active levels. The environmental persistence of several commonly used medicinal drugs, such as erythromycin, cyclophosphamide, naproxen, sulphamethoxazole, or sulphasalazine, is longer than one year [Zuccato et al. 2004, Castiglioni et al. 2004b, Zuccato et al. 2006]. Clofibric acid, the main metabolite of clofibrate, a drug used to lower cholesterol, is still detectable in rivers years after its withdrawal from the market [Zuccato et al. 2000]. Several pharmaceuticals are consequently detectable in surface waters (rivers, lakes and seas) in the ng/L up to the μg/L range [Calamari et al. 2003, Zuccato 2004, Castiglioni et al. 2006]. The recent literature is therefore rich in studies on levels of pharmaceuticals in the environment. However, thousands of different pharmaceuticals are currently used worldwide, with hundreds of new molecules synthesized every year to replace obsolete compounds. Moreover because of differences in the prevalence of diseases, treatment habits and options, or simply for market reasons, the pollution profile can vary significantly among different countries. Contamination by pharmaceuticals is therefore a variable and changing issue, needing continuous monitoring.

Occurrence of Pharmaceuticals in Surface, Waste, Drinking Water and Sediments

Except for two preliminary studies that measured levels of pharmaceuticals in the environment, published in 1977 [Hignite andAzarnoff 1977] and in 1985 [Richardson and Bowron 1985], the first work with a systematic approach started to appear in the 1990s, dealing with levels of pharmaceuticals in local rivers, groundwater and STPs in Germany [Ternes 1998, Hirsch et al. 1999,Heberer et al. 2002]. The presence of several pharmaceuticals in STP effluents was soon confirmed in The Netherlands [Belfroid et al. 1999], Switzerland [Golet et al. 2001], United Kingdom [Johnson and Sumpter 2001], France, Greece, and Sweden [Andreozzi et al. 2003], Spain [Carballa et al. 2004] USA [Kolpin et al. 2002, Huggett et al. 2003, Yang and Karlson 2004], Canada [Metcalfe et al. 2003], Brazil [Stumpf et al. 1999] and Australia [Khan and Ongerth 2005, Braga et al. 2005]. Pharmaceuticals have also been reported in measurable concentrations in sewage sludge, indicating that alternative routes to the environment may also be significant [Khan and Ongerth 2002].

In Italy, several "priority" pharmaceuticals in water and sediments from the rivers Po, Lambro and Adda were reported [Zuccato et al. 2000]. Drinking water samples from the mains water systems showed traces of diazepam, clofibric acid and tylosin in samples from the cities of Lodi and Varese. Traces of some pharmaceuticals were also present in sediments from the same rivers. The widespread occurrence of pharmaceuticals in the Italian environment was subsequently confirmed by the same authors [Calamari et al. 2003], and by others [Andreozzi et al. 2003]. The former paper, [Calamari et al. 2003], reports results of an extensive investigation with eight sampling stations along the rivers Po and Lambro, from which were plotted the patterns of contamination in a highly populated region with a large number of animal farms. In this study, 18 pharmaceuticals could be detected in significant amounts. Further studies subsequently measured pharmaceuticals for human use in the influents and effluents of urban STPs across the whole Italy, where several drugs were found in the microgram per liter range. By comparing STP levels with those previously measured in rivers, it was possible to identify a group of pollutants important for the environment in Northern Italy [Zuccato et al. 2005a]. These pharmaceuticals were detected in high concentrations in STP effluents and were also persistent enough to remain in substantial amounts in river waters.

A number of studies have identified APIs in potable water sources in the UK, Canada, Germany, France and, in particular, in USA where carbamazepine and several antibiotics (fluoroquinolones, sulfonamides and macrolides) have been detected at ng/L concentrations [Bruchet et al. 2005, Zuhlke et al. 2004, Stackelberg et al. 2004, Ternes and Joss, 2006]. More recently, a survey of contamination of surface and drinking waters around Lake Maggiore in Northern Italy reports that several drugs, including carbamazepine, sulfamethoxazole, gemfibrozil, carbamazepine and benzafibrate, were found in potable water at almost identical levels with those found in the lake itself [Loos et al. 2007]. Similarly, Rabiet and colleagues studied the consequences of treated water re-use in surface and ground waters, detecting pharmaceuticals and other wastewater-related contaminants in several reservoirs tapped for drinking water, confirming that drinking water contamination was frequently associated with paracetamol, caffeine, and diclofenac [Rabiet et al. 2006].

Removal by Conventional Sewage Treatment Processes

Hundreds of tons of pharmacologically active substances can enter STPs each year [Khan and Ongerth 2004]. For many drugs, removal by conventional biological treatments seems inefficient [Ternes 1998, Stumpf et al. 1999, Metcalfe et al. 2003, Carballa et al. 2004, Khan and Ongerth 2005], since they are found in measurable quantities in STP effluents and surface water [Hirsch et al. 1999, Calamari et al. 2003]. STPs might therefore be important point sources of contamination, but for the majority of pharmaceuticals little information is available on their behaviour and ultimate fate in STPs and in receiving surface waters. The removal rate of pharmaceuticals in STPs is potentially affected by several factors, such as the nature of the drug, the treatment process employed, the age of the activated sludge, environmental conditions such as the season, and the characteristics of the influent [O'Brien and Dietrich 2004].

A recent study used mass balance calculations to investigate six STPs in Italy to gain further knowledge of the removal and behaviour of drugs during treatment. Removal efficiencies were assessed for 30 compounds and the factors affecting removal in STPs were identified [Castiglioni et al. 2006]. Considering the total loads, removal rates were generally less than 40%, with the exception of a plant in Varese Lago, where it was 64% (average removal for all tested pharmaceuticals). Apparent seasonal differences were observed with improved removals duing summer months. This study found that the removal rates of pharmaceuticals in STPs can vary significantly over a year, but abatement is typically incomplete. In addition to long-term seasonal variations, short-term diurnal variations may be significant for some pharmaceuticals, indicating that care must be taken to determine representative sampling regimes [Khan and Ongerth 2005].

Prioritisation of Pharmaceuticals

Environmental contamination by pharmaceuticals is an emerging concern, warranting monitoring for several reasons. These include the need for reliable assessment of environmental and human health risks according to variable exposure scenarios. However, the establishment of meaningful monitoring programs is complicated by the shear number and diversity of active constituents in use, many with different chemical structures and physico-chemical properties. It is therefore necessary to develop a means of prioritisation so as to restrict studies to a limited number of hazardous molecules. The recent literature provides some examples of efforts to enact such prioritisation. One method used in the USA for preliminary screening took into account the number of prescriptions and sales figures [Erickson 2002]. Another approach also considered the wastewater pathways, the human and environmental implications, the representativity of certain classes of compounds or sources and the availability of analytical methods to detect the compounds [Kolpin et al. 2002]. In Germany, analysis was limited to pharmaceuticals produced in amounts over ten tons/year, for which analytical methods and reference standards were available [Erickson 2002].

As an example, a group of potential contaminants was identified in Italy, selected on the basis of their mass balance and theoretical loads, i.e. multiplying sales by percentages of metabolic excretion in humans [Zuccato et al. 2000, Calamari et al. 2003]. This provided a list of "top" pharmaceuticals for human use, excreted unmetabolised in substantial amounts, with estimated environmental loads in the order of tons per year. This approach has been developed further by consideration of relative proportions of specific compounds which are excreted unmetabolised and modelled behaviour during conventional sewage treatment processes [Khan and Ongerth 2004]. Other groups of pharmaceuticals have been selected based on the fact that they have been previously reported in the environment [Zuccato et al. 2000]. Several new molecules, however, are synthesised every year by the pharmaceutical industry, so contamination is a changing issue, needing continuous updating [Castiglioni et al. 2004b, Zuccato et al. 2004]. Table 1 shows the updated list of "priority pharmaceuticals" for monitoring and ecotoxicology studies, based on our published work and on recent literature data.

The process of prioritisation of pharmaceuticals is also crucial for the toxicological evaluation of drug mixtures. Aquatic organisms and humans are exposed simultaneously or sequentially to large numbers of chemicals via multiple exposure routes. Determining health risks from exposure to chemical mixtures is a daunting challenge to many experimental toxicologists and epidemiologists using observational methods. Different types of mixtures require different approaches, and the usefulness of a certain approach depends on the amount, type and quality of the available data on the chemistry and the toxicity of the mixture. A comprehensive scheme for the safety evaluation of complex mixtures has been developed some years ago [Groten et al. 2001, Feron and Groten 2002]. This scheme is summarised in a slightly modified form in Figure 1. The most conspicuous element of this scheme is the reduction of complex mixtures into mixtures that are readily available for testing in their entirety, using the "top n" approach. The "top n" approach is similar to the one describe in the previous paragraph, and means identification of the n (for example, the 10) most risky (not most toxic) chemicals in a complex mixture. Once the "top n" chemicals have been identified, they will be approached as a simple (defined) mixture, assuming that hazard and potential risk of the mixture of the "top n" chemicals are representative for the hazard and risk of the entire mixture. A very similar approach is successfully being used in the United States to identify the priority substances that are released from hazardous waste sites and that pose the greatest risk to human health [Johnson and DeRosa, 1995]. We found that the "top n" approach seems to be particularly suitable for the toxicological evaluation of complex drug mixtures, as evident in the following sections dealing with ecotoxicology and human health [Pomati et al. 2006, Pomati et al. 2007].

Table 1. Pharmaceuticals of priority concern for surface and ground waters

Pharmaceutical	Therapeutic category	Reference
Atenolol	Anti-hypertensive	Calamari et al. 2003
Bezafibrate	Lipid regulating	Calamari et al. 2003
Carbamazepine	Anticonvulsant/antidepressant	Castiglioni et al. 2006
Ciprofloxacin	Antibacterial (fluoroquinolone)	Calamari et al. 2003
Cyclophosphamide	Cytotoxic (alkylating anticancer agent)	Zuccato et al. 2000
Diclofenac	Nonsteroidal antiinflammatory	Rabiet et al. 2006
Furosemide	Diuretic	Calamari et al. 2003
Gemfibrozil	Lipid regulating	Loos et al. 2007
Hydrochlorothiazide	Diuretic	Calamari et al. 2003
Ibuprofen	Nonsteroidal antiinflammatory	Zuccato et al. 2000
Lincomycin	Antibacterial (lincosamine)	Calamari et al. 2003
Ofloxacin	Antibacterial (fluoroquinolone)	Castiglioni et al. 2004-b
Ranitidine	Ulcer healing	Calamari et al. 2003
Paracetamol	Nonsteroidal antiinflammatory	Rabiet et al. 2006
Salbutamol	Bronchodilator	Zuccato et al. 2000
Sulfamethoxazole	Antibacterial (sulfonamide)	Castiglioni et al. 2004-b

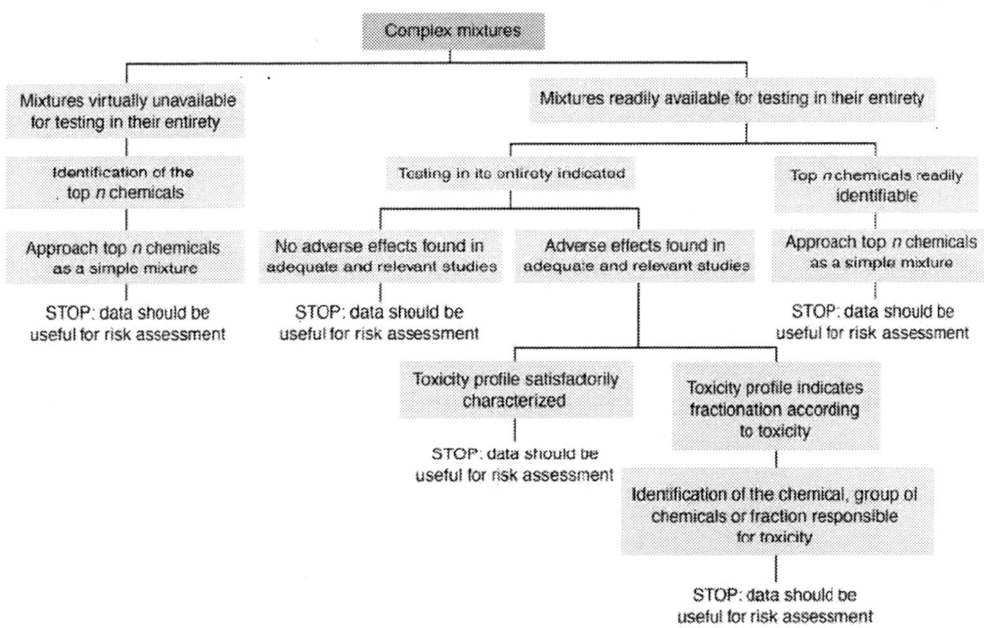

Figure 1. Safety evaluation of complex mixtures as proposed by Groten et al. 2001, with slight modifications. In this framework, simple mixtures can be tested entirely and/or dismantled, testing individual components and their combinations to study additivity and interaction.

ECOTOXICOLOGY AND ENVIRONMENTAL HEALTH

The widespread occurrence of pharmaceuticals in aquatic habitats has raised a novel and intricate environmental issue. The potential risk associated with the presence of these contaminants in surface waters is just beginning to be understood. The effects of APIs on aquatic life have been largely investigated [for a comprehensive review see Fent et al. 2006, and Crane et al. 2006]. Aquatic toxicology data for human and veterinary medicines, however, are scattered across studies targeting a range of different test organisms, in different conditions, often with incoherent results. The current lack of knowledge holds in particular for chronic effects, which have rarely been investigated. Toxicity studies in this field have to face a number of difficulties. Drugs are present in the environment at very low concentrations, far below the doses employed for medical treatment, so acute toxicity tests do not detect any biological effects [Boxall et al. 2003]. Target and non-target organisms are therefore exposed to low levels of pharmaceuticals in the environment, but for extended periods of time - sometimes permanently.

A potential effect of pharmaceuticals in the environment that may have repercussions on aquatic life and/or humans is related to the presence of antibiotics in waste, surface and ground waters. Exposure to antibiotics might induce antimicrobial resistance [Kummerer 2004b], and horizontal transfer of resistance genes in field bacterial populations [Davidson 1999]. Persistence of bacteria resistant to antibiotics, despite a decrease in antibiotic prescriptions, has been reported for sulphonamide in the UK [Enne et al. 2001]. However, the ban of antibiotics as growth promoters in animal farming, such as for vancomycin in Sweden (1986) and for avoparcin in Denmark (1995), resulted in a significant decrease of resistance in manure and enteric bacteria [Van de Bogaard and Stobberingh 1999]. Recently, Cunha [2001] suggested that the development of antibiotic resistance might not be related to the total amounts of antibiotics used, but to the potency of the single compounds. We recently investigated the correlation between the presence of antibiotics in the environment and the development of resistance in bacteria, targeting the multidrug-resistance gene marA. marA is induced by tetracycline, chloramphenicol, ampicillin, nalidixic acid and ciprofloxacin, and its expression is sufficient to produce a multidrug resistance phenotype [Goldman et al. 1996]. Samples of sediment and water from rivers and treatment plants showed the ubiquitous presence of marA, with plated samples giving rise to antibiotic resistant colonies, including *Bacillus* strains that appeared to have horizontally acquired the gene form enterobacteria in sewage [Castiglioni et al. 2005b]. Our quantitative data, however, suggest that resistance is ubiquitous in the environment and therefore not directly related to the presence/absence of water-borne antibiotics. This confirms that the environment is a major reservoir for antibiotic resistance [Biyela et al. 2004]. Antibiotic local concentrations, however, determine the amount and distribution of antibiotic-resistance genes present in the endemic bacterial population [Castiglioni et al. 2005b].

Therapeutic drugs are ordinarily found in complex mixtures of unrelated chemicals with a range of pharmacological activities, suggesting the potential for additive and synergistic effects [Cleuvers 2003, Pomati et al. 2006a]. This raises questions about the possible danger posed by water-borne drugs on the survival and physiology of aquatic life, exposed to the bioactive chemicals for all their life cycle, or the ability of certain APIs to induce genetic alterations or mutations. Pharmaceuticals in the environment are suggested to pose only a low risk for acute toxicity. For chronic effects, the situation may be different, but there is a considerable lack of information. Investigation of multigenerational life-cycle effects or at various life stages is lacking, although many aquatic organisms are exposed for their entire life. There is a need to focus on long-term exposure assessment regarding specific modes of action of pharmaceuticals to better judge the implications of pharmaceutical residues in aquatic systems. Only after filling these gaps, more reliable environmental risk assessments with much lower uncertainty can be performed.

It appears from several studies that algae and cyanobacteria are the most sensitive models with which to study short and long-term effects of pharmaceuticals [Fent et al. 2006, Crane et al. 2006], including cocaine-like local anaesthetics [Suzuki et al. 2004]. A possible promotion of cyanobacterial blooms by certain APIs in contaminated surface waters has been suggested [Pomati et al. 2003]. In cyanobacterial species common in freshwaters, local anaesthetics have been shown to trigger the production of the potent neurotoxin saxitoxin, and induce the expression of genes associated with bloom development and toxic-scum formation [Pomati et al. 2006b]. Higher aquatic plants such as duckweed have been demonstrated to be more responsive to certain human pharmaceuticals as compared to green algae and microalgae [Pomati et al. 2004, Crane et al. 2006]. Among the animals, aquatic invertebrate species showed adverse effects at concentrations of APIs closer to real environmental exposure scenarios [Fent et al. 2006, Crane et al. 2006]. Conversely, there is little evidence for any direct adverse effects of pharmaceuticals on aquatic vertebrates, including amphibians and fish, at environmentally realistic concentrations [Crane et al. 2006].

For assessing the acute toxicity of pharmaceuticals, *in vitro* systems are rarely considered. This is also true for chronic toxicity parameters that may be important for long-term effects. Caminada and colleagues [2006] evaluated the *in vitro* cytotoxicity of 34 common pharmaceuticals from different classes and with different modes of action using two fish cell lines. Cytotoxicity was found for 21 pharmaceuticals with EC50-values in the mg/L range. Recently, we studied the effects of a mixture of pharmaceuticals at environmentally relevant concentrations on the growth and transcriptional regulation in zebrafish liver (ZFL) cells [Pomati et al. 2007]. APIs of interest were selected among those detected in northern Italian rivers [Calamari et al. 2003, Castiglioni et al. 2004b] and prioritised based on occurrence in the environment as well as on long-term use forecast, poor biodegradation, poor metabolization, and high sale levels [Pomati et al. 2006a]. The consequent list of top pharmaceuticals causing potential environmental risk comprised atenolol, bezafibrate, carbamazepine, cyclophosphamide, ciprofloxacin, furosemide, hydrochlorothiazide, ibuprofen, lincomycin, ofloxacin, ranitidine, salbutamol, and sulphamethoxazole. This mixture of 13 APIs was assembled to mimic the association and low concentrations (ng/L) of drugs as detected in the environment, and decidedly inhibited ZFL cells proliferation *in vitro* over a 72 h exposure time. Using an oligonucleotide DNA-microarray targeting 14,000 zebrafish transcripts, we profiled gene expression in ZFL cells treated with ecologically relevant levels of the drug mixture, observing changes in transcripts that could be associated with general toxicity, and others that were possibly linked to the effects of specific responses, including endocrine disruption. We also found a similarity in the response to the drug mixture in fish and previously employed human cells [Pomati et al. 2006a]. Several fish cell lines from different species and different organs have been previously used for cytotoxicity assays [Fent, 2001], and they have been shown to estimate the basal toxicity equally as well as mammalian cells when correlating the EC50 values [Castano and Gomez-Lechon, 2005].

A critical step towards the use of *in vitro* assays as models for *in vivo* animal experiments is the correlation between *in vitro* and *in vivo* activities [Leusch et al. 2007]. A significant correlation between the *in vitro* toxocity and *in vivo* toxicity was found for *Daphnia* by Caminada et al. [2006], but not for fish due to insufficient and heterogeneous data. Our preliminary work provides an additional indication that *in vitro* toxicological data suggest observed effect concentration values comparable with standard chronic exposure tests using aquatic invertebrates. Overall, we propose that *in vitro* cytotoxicity assays with cell lines could be suited for the first screening of *in vivo* toxicity of pharmaceuticals, thereby contributing to the reduction of *in vivo* experiments.

Drinking Water Treatment and Advanced Water Treatment Processes

Conventional drinking water treatment processes such as coagulation or chemical lime softening appear to be only marginally effective for the removal of most pharmaceuticals [Shon et al. 2006, Westerhoff et al. 2005].

Powdered activated carbon (PAC) has been shown to be highly effective for the removal of a wide range of pharmaceuticals from relatively clean water sources [Adams et al. 2002, Westerhoff et al. 2005]. A study recently undertaken at the Southern Nevada Water Authority provides a useful illustrative example [Westerhoff et al. 2005]. For this research, raw drinking water supplies were collected and high concentrations of 62 different chemicals were spiked into them. These waters were then treated by a number of laboratory-scale water treatment processes including PAC. Addition of 5 mg/l of PAC with a 4 hour contact time removed different compounds by between 10% to greater than 98%. Higher PAC dosages improved the removal of most compounds. This study confirmed that the removal effectiveness for specific chemicals could be reasonably well predicted based on their lipophilicity.

Granular activated carbon (GAC) has also been shown to be effective for the removal for some important organic chemical contaminants in water. For example, Ternes et al. [2002] investigated the removal of some pharmaceuticals during a range of drinking water treatment processes. The particular pharmaceuticals were some that are commonly reported in European drinking water sources including carbamazepine, diclofenac, clofibric acid, bezafibrate and primidone. This study revealed GAC filtration to be an effective method for removing most of the studied compounds. It was shown that GAC treatment, in combination with other conventional treatment processes, significantly removed most of the pharmaceuticals.

Advanced oxidation processes (AOPs) can be used for the mitigation of most organic chemicals in water, including pharmaceuticals. The overall extent of oxidation for any AOP is dependant on the contact time and the concentration of scavengers in the water (ie non-target oxidisable species). Typically, dissolved organic carbon (DOC) and carbonate/bicarbonate are the most important scavengers in drinking waters. High concentrations of DOC and carbonate/bicarbonate can render mineralisation of micropollutants quite inefficient and very costly [von Gunten 2003]. However, pre-treatment processes such as GAC or RO significantly reduce DOC concentrations, thus enhancing oxidation efficiency.

Oxidation by direct application of molecular ozone is an effective process for some contaminants. For example, many pharmaceuticals can be oxidised to more than 90-99%

using typical ozone treatment doses [Huber et al. 2005, Irmak et al. 2005, Ternes et al. 2003, Westerhoff et al. 2005]. Typical doses depend on the initial water quality, but are normally calculated to achieve a residual of 0.2-0.3 mg/L after 3 min of contact time and zero residual within 10 min [von Gunten 2003, Westerhoff et al. 2005]. However, advanced oxidation utilising hydrogen peroxide is a more effective process for an even wider range of these target species [Huber et al. 2003, von Gunten 2003, Zwiener and Frimmel 2000].

Investigations on some active pharmaceuticals such as the contraceptive hormone, ethynylestradiol [Huber et al. 2004], and the anti-epileptic drug, carbamazepine [McDowell et al. 2005], have shown that even partial oxidation is sufficient to reduce pharmacological activity and toxicity of these agents.

A study of pharmaceutical and estrogenic hormone removal was undertaken at an advanced water recycling demonstration plant in Australia [Khan et al. 2004]. This study involved spiking unnaturally high concentrations of the selected chemicals into the influents of various stages of the treatment train to test performance. In the case of reverse osmosis, a single membrane module was used and was shown to be the most effective barrier for the removal of the investigated compounds. Studies of these chemicals in full-scale operational reverse osmosis plants are limited due to the fact that concentrations are generally already very low prior to membrane treatment and such 'spiking' experiments are unlikely to be permitted by regulatory authorities or plant operators. However, a recent study undertaken at two full-scale reverse osmosis plants in the USA identified numerous pharmaceuticals in the RO feed waters [Drewes et al. 2005]. The permeate water from this plant did not reveal any quantifiable detects except for low concentrations of caffeine at one facility. In other words, the pharmaceuticals were completely removed by the RO membrane, as best as the sensitivity of the analytical method could determine.

Human Health Risk Associated with Water-Borne Pharmaceuticals

The effects of trace and mixed contaminants of pharmaceutical origin on humans, influencing water for drinking and other health related activities, have been virtually ignored until very recently [Weber et al. 2006]. Surface water is widely used for drinking water and thus, the occurrence of APIs in surface waters poses potential health issues to society in general and water utilities in particular. Evidence has been provided that a mixture of APIs at environmentally relevant levels can affect the physiology of human cells [Pomati et al. 2006a]. However, there has not been any direct proof that very low concentrations of APIs have adverse health effects on humans. Nonetheless, it would be premature to assume that such evidence will not be forthcoming from future investigations. Low levels of endocrine disruptors (EDCs) in water are an increasing concern for society because EDCs exert physiological effects at very low concentrations that may translate into human impairment. Precautionary thinking ordains that drinking water should be free from trace APIs to minimise the unpredictable long-term risks. In the face of uncertainty about the consequences of continuous exposure of the population to trace quantities of APIs, improved understanding of the potential implications of very low concentrations of APIs in drinking water is required.

Pharmaceuticals are designed to stimulate or inhibit specific physiological responses. Continuous releases and chronic exposure can result in subtle effects on aquatic species and could pose a risk to human health associated with consuming contaminated drinking water over a lifetime [Dorne et al. 2007, Khetan and Collins 2007]. There is scant information in the literature addressing the potential effects to human health from the presence of trace levels of pharmaceuticals found in surface and drinking water [Schwab et al. 2005, Cunningham et al. 2006]. It is widely recognised, however, that chronic effects arising from long-term exposure to some pharmaceuticals at much lower concentrations may follow different toxicodynamic mechanisms than extrapolated from short-term acute studies [Lange and Dietrich 2002]. Additionally, individual pharmaceuticals occur in the environment in combination with other pharmaceutical compounds together with their own metabolites and degradation by-products, as well as other environmental pollutants [Escher et al. 2005]. As current environmental risk assessments tend to focus on single substances, it is likely that the prevailing assessments underestimate the real environmental impacts, and some examples of this have been discussed above. Thus, cumulative exposures need be considered for pharmaceuticals. This should clearly entail mixtures of APIs that share a common mode of toxic action, as well as mixtures with diverse modes of action.

In previous studies, a mixture of drugs was studied *in vitro* and significantly inhibited the growth of human embryonic cells and zebra fish liver cells, affecting their physiology and morphology [Pomati et al. 2006, Pomati et al. 2007]. In human and fish cells, exposure to drugs at environmental levels disrupted the regulation of conserved molecular mechanisms, including signal transduction, genes related to the progression of the cell-cycle, primary and lipid metabolism, as well as stress and immune response. Many of those genes also play a role in embryo development. An increased expression of genes that are characteristic of cancer was observed in fish cells treated with APIs [Pomati et al. 2007].

We therefore aimed at the identification of the APIs in the "model" mixture that were responsible for the observed biological effects. To achieve this goal, we planned to limit our mixture of 13 drugs down to chemicals having a direct effect on cell proliferation only, or being able to interact with other compounds with this respect. We chose *in vitro* cell proliferation as the endpoint and explored several factorial experiments, based on a statistically designed approach, to investigate principal effects and possible interactions occurring among the chemicals composing our mixture at environmentally relevant levels. We found a high intrinsic variability of *in vitro* toxicity tests in preliminary analyses, and selected randomised fractional factorial designs as the optimal solution for our objectives. We studied the effects of tens of different combinations of APIs on human cell growth (Pomati et al., unpublished). Initial data show that only a few drugs, including antibiotics, have a reproducible and statistically significant effect on cell proliferation. Moreover, evidence was found for interactions occurring among drugs in the mixture at environmental concentrations (Pomati et al., unpublished). These studies represented the primary substantial evidence that a potential risk is associated with the presence of environmental pollutants of mixed pharmaceutical origin.

As described by the US EPA, safe concentrations of environmental pollutants are typically determined with reference to an 'acceptable daily intake' (ADI) value [US Environmental Protection Agency Risk Assessment Forum 2002] . It is common for many chemicals have several known toxic endpoints and the ADI is normally determined from the most restrictive (lowest) of them. For most pharmaceutical agents, the therapeutic effect normally occurs at a dose considerably below those expected to cause toxicity, thus the lowest therapeutic dose is commonly equivalent to the Lowest Observed Effect Level (LOEL) [Schwab et al. 2005]. An appropriate uncertainty factor can then be applied to estimate a No Observed Effect Level (NOEL) from a LOEL as part of the ADI calculation.

There are currently no mandated guidelines for the management of pharmaceutical residuals, which may be present in drinking water as a result of planned or unplanned indirect potable water recycling. However, the first efforts towards a regulatory approach for establishing such guidelines has recently been published in the draft Phase 2 Australian Guidelines for Water Recycling [Australian Guidelines for Water Recycling 2007]. These draft guidelines specifically relate to the augmentation of drinking water supplies with recycled water and include a strong focus on pharmaceutical residuals.

The guidelines take advantage of the fact that the biological or pharmacological activity at therapeutic doses for pharmaceuticals used for humans is well known, and can be found in the manufacturer's literature and in various pharmacopoeias. The approach taken to establish guideline concentrations for pharmaceuticals in Australia was to assume that a reasonably confident NOEL could be determined by applying a suitable safety factor to a LOEL, which in turn was assumed to be equivalent to the lowest recommended therapeutic dose [Australian Guidelines for Water Recycling 2007]. These assumptions appear to be supported for a majority of considered pharmaceuticals [Schwab et al. 2005]. However, it remains likely that numerous exceptions will be identified, which may require individual consideration of specific toxicity data. These may include prescription drugs for which consumption, even at acute NOELs, may have long-term exposure impacts, with particular risks to pregnant women and infants.

Conclusions and Future Research Needs

Pharmaceuticals in the environment are increasingly a matter of global concern, with potential consequences for humans and ecosystems. However, the effects of life-long exposures are still to be determined. For instance, we do not know whether the presence of antibiotic drugs in the environment contributes to bacterial resistance, or to what extent drugs can influence human health or be transferred to humans through food-chain biomagnification. Further research must be encouraged, and the human and environmental implications of the discharge of APIs into the environment requires thorough evaluation. Knowledge is rapidly growing, but the extent of potential impacts is not yet fully clear, and current information on risk assessment and management is far from adequate. Increased knowledge of the causes, occurrence and effects of APIs as environmental pollutants is required for improved understanding of this recently emerging issue to improve abatement strategies, and to mitigate environmental consequences.

APIs are ordinarily detected in complex mixtures of unrelated molecules, and interactions between classes of drugs represent a serious toxicological issue, still also largely unresolved in medicine. Furthermore, APIs may have chronic impacts to a variety of organisms in different stages of development and for which no pharmacological action is known. Further concerns regard the capability of complex APIs mixtures for inducing genetic alteration/mutation, or the possible risks posed by waterborne APIs in water recycling for human use. An effective and sustainable strategy against this insidious and mostly unseen contamination of aquatic environments barely exists. Information on API behaviour in the environment is crucial to inform a reliable approach to risk assessment, and to identify appropriate remediation and control measures. Here we propose three crucial research needs that we identify as a priority in this field:

- A constant development of monitoring programmes and techniques to survey water resources. Initially it is necessary to improve knowledge regarding which particular APIs persist in the environment under certain conditions. This information can be used in the longer term to inform monitoring programs to better target specific APIs, which may act as indicator compounds and thus provide a greater scope of information regarding the likely presence of a wider range of compounds for specific situations.
- Standardised and high throughput toxicology data, in particular to predict long term exposure effects and study mixture toxicity. Simple aquatic organisms such as invertebrates are good toxicological models to study, being the most sensitive part of water ecosystems, and being potential sentinels for effects on the most sensitive human archetype: the foetus.
- Identification of remediation and mitigation strategies in water treatment to reduce the potential hazards, implementing the risk management process of ensuring that populations and individuals are not chronically exposed to hazardous concentrations of chemical contaminants in water.

References

Adams, C, Wang, Y, Loftin, K, and Meyer, M (2002) Removal of antibiotics from surface and distilled water in conventional water treatment processes. *J. Environ. Eng.-ASCE.* 128, 253-260.

Andreozzi R, Marotta R, and Nicklas P (2003) Pharmaceuticals in STP effluents and their solar photodegradation in aquatic environment. *Chemosphere* 50, 1319-1330.

Anway, MD, Cupp, AS, Uzumcu, M, and Skinner, MK (2005) Epigenetic transgenerational actions of endocrine disruptors and male fertility. *Science* 308, 1466-1469.

Australian Guidelines for Water Recycling: Managing Health and Environmental Risks (Phase 2): Augmentation of Drinking Water Supplies (2007) National Water Quality Management Strategy – Natural Resource Management Ministerial Council and Environment Protection and Heritage Council Publication, ISBN 1 921173 20 3 (http://www.ephc. gov.au/ephc/water_ recycling.html).

Belfroid. AC, Van der Horst. A, Vethaak. AD, Schafer. AJ, Rijs. GBJ, Wegener. J, and Cofino, WP (1999) Analysis and occurrence of estrogenic hormones and their glucuronides in surface water and wastewater in The Netherlands. *Sci. Total. Environ.* 225, 101-108

Biyela, PT, Lin, J, Bezuidenhout, CC (2004) The role of aquatic ecosystems as reservoirs of antibiotic resistant bacteria and antibiotic resistance genes. *Water Sci. Technol.* 50, 5-50.

Boxall, AB, Kolpin, DW, Halling-Sorensen, B, and Tolls, J (2003) Are veterinary medicines causing environmental risks? *Environ. Sci. Technol.* 37, 286A–294A.

Braga, O, Smythe, GA, Schafer, A, and Feitz AJ (2005) Fate of steroid estrogens in Australian inland and coastal wastewater treatment plants. *Environ. Sci. Technol.* 39, 3351- 3358.

Bruchet, A, Hochereau, C, Picard, C, Decottignies, V, Rodrigues, JM, and Janex-Habibi, ML (2005) Analysis of drugs and personal care products in French source and drinking waters: the analytical challenge and examples of application. *Water Sci. Technol.* 52, 53-61.

Calamari, D, Zuccato, E, Castiglioni, S, Bagnati, R, and Fanelli, R (2003) Strategic survey of therapeutic drugs in the rivers Po and Lambro in northern Italy. *Environ Sci Technol* 37, 1241-1248.

Caminada, D, Escher, C, and Fent, K (2006) Cytotoxicity of pharmaceuticals found in aquatic systems: Comparison of PLHC-1 and RTG-2 fish cell lines. *Aquatic Toxicol.* 79, 114–123.

Carballa, M, Omil, F, Lema, JM, Llompart, M, Garcia-Jares, C, Rodriguez, I, Gomez, M, and Ternes, T (2004) Behavior of pharmaceuticals, cosmetics and hormones in a sewage treatment plant. *Water Res* .38, 2918-2926.

Castano, A and Gomez-Lechon, MJ (2005) Comparison of basal cytotoxicity data between mammalian and fish cell lines: a literature survey. *Toxicol. In Vitr* .19, 695–705.

Castiglioni, S, Fanelli, R, Calamari, D, Bagnati, R, Zuccato, E. (2004b) Methodological approaches for studying pharmaceuticals in the environment by comparing predicted and measured concentrations in River Po, Italy. *Regul Toxicol. Pharmacol. 39*, 25-32.

Castiglioni, S, Pomati, F, Miller, K, Zuccato, E, Calamari, D, and Neilan BA (2005b): Presence of antibiotic-resistance genes in the environment. SITE (Società Italiana di Ecologia), XV annual meeting, 12-14 September, Torino.

Castiglioni, S, Bagnati, R, Fanelli, R, Pomati, F, Calamari, D, and Zuccato, E (2006) Removal of pharmaceuticals in sewage treatment plants in Italy. Environ Sci Technol 40, 357-363.

Cleuvers, M (2003) Aquatic ecotoxicity of pharmaceuticals including the assessment of combination effects. *Toxicol. Lett.* 142, 185-194.

Crane, M, Watts, C, and Boucard, T (2006) Chronic aquatic environmental risks from exposure to human pharmaceuticals. *Science of the Total Environment.* 367, 23–41.

Cunha, BA (2001) Effective antibiotic-resistance control strategies. *The Lancet* 357, 1307-1308.

Cunningham, VL, Buzby, M, Hutchinson, T, Mastrocco, F, Parke, N, and Roden, N (2006) Effects of human pharmaceuticals on aquatic life: next steps. Environ .Sci. Technol. 40, 3456.

Davidson, J (1999) Genetic exchange between bacteria in the environment. *Plasmid* 42, 73-91.

Dorne, J, Skinner, L, Frampton, GK, Spurgeon, DJ, and Ragas, AMJ (2007) Human and environmental risk assessment of pharmaceuticals: differences, similarities, lessons from toxicology. *Anal. Bioanal. Chem.* 387, 1259.

Drewes, JE, Bellona, C, Oedekoven, M, Xu, P, Kim, U, and Amy, G (2005) Rejection of wastewater-derived micropollutants in high-pressure membrane applications leading to indirect potable reuse. *Environ. Progress* 24, 400-409.

Enne, V, Livermore, D, Stephens, P, and Hall, L (2001) Persistence of sulphonamide resistance in Escherichia coli in the UK despite national prescribing restriction. *The Lancet* 357, 1325-1328.

Environmental Protection Agency Risk Assessment Forum (US). A Review of the Reference Dose and Reference Concentration Processes. Washington, DC; 2002

Erickson, BE (2002) Analyzing the ignored environmental contaminants. *Environ. Sci. Technol.* 36, 140A-145A.

Escher, BI, Bramaz, N, Eggen, RIL, and Richter, M (2005) In vitro assessment of modes of toxic action of pharmaceuticals in aquatic life. *Environ. Sci. Technol.* 39, 3090.

Fent, K (2001) Fish cell lines as versatile tools in ecotoxicology: assessment of cytotoxicity, cytochrome P4501A induction potential and estrogenic activity of chemicals and environmental samples. *Toxicol. In Vitr.* 15, 477–488.

Fent, K, Weston, AA, and Caminada, D (2006) Ecotoxicology of human pharmaceuticals. *Aquatic. Toxicol.* 76, 122–159.

Feron, VJ and Groten, JP (2002) Toxicological evaluation of chemical mixtures. *Food Chem Toxicol* 40, 825–839.

Goldman, J, White, D, Levy, S (1996) Multiple antibiotic resistance (mar) locus protects Escherichia coli from rapid cell killing by fluoroquinolones. *Antim. Agents Chemot.* 40, 1276-1269.

Golet, EM, Alder, AC, Hartmann, A, Ternes, TA, and Giger, W (2001) Trace determination of fluoroquinolone antibacterial agents in urban wastewater by solid-phase extraction and liquid chromatography with fluorescence detection. *Anal. Chem.* 73, 3632-3638.

Groten, JP, Feron, VJ, and Sühnel, J (2001) Toxicology of simple and complex mixtures. *Trends Pharmacol. Sci.* 22, 316-322.

Heberer, T (2002) Occurrence, fate, and removal of pharmaceuticals residues in the aquatic environment: a review of recent research data. *Toxicol. Lett.* 131, 5-17.

Hignite, C and Azarnoff D (1977) Drugs and drug metabolites as environmental contaminants: chlorophenoxyisobutyrate and salicyclic acid in sewage water effluent. *Life Sci.* 20, 337-341.

Hirsch, R, Ternes, T, Haberer, K, and Kratz, L (1999) Occurrence of antibiotics in the aquatic environments. *Sci. Total Environ.* 225, 109-118.

Huber, MM, Canonica, S, Park, GY, and von Gunten, U (2003) Oxidation of pharmaceuticals during ozonation and advanced oxidation processes. *Environ. Sci. Technol.* 37, 1016-1024.

Huber, MM, Ternes, TA, and von Gunten, U (2004) Removal of estrogenic activity and formation of oxidation products during ozonation of 17 alpha-ethinylestradiol. *Environ. Sci. Technol.* 38, 5177-5186.

Huber, MM, Gobel, A, Joss, A, Hermann, N, Loffler, D, and McArdell, CS (2005) Oxidation of pharmaceuticals during ozonation of municipal wastewater effluents: A pilot study. *Environ. Sci. Technol.* 39:4290-4299.

Huggett, DB; Khan, IA, Foran, CM, and Schlenk, D (2003) Determination of beta-adrenergic receptor blocking pharmaceuticals in United States wastewater effluent. *Environ. Poll.* 121, 199-205.

Irmak, S, Erbatur, O, and Akgerman, A (2005) Degradation of 17 beta-estradiol and bisphenol A in aqueous medium by using ozone and ozone/UV techniques. *J. Haz. Mat.* 126, 54-62.

Johnson, BL and DeRosa, CT (1995) Chemical mixtures released from waste sites: implications for health risk assessment. *Toxicol.* 105, 145–156.

Johnson AC, and Sumpter JP (2001) Removal of endocrine-disrupting chemicals in activated sludge treatment works. *Environ. Sci. Technol.* 35, 4697-4703.

Khan, SJ and Ongerth, JE (2002) Estimation of pharmaceutical residues in primary and secondary sewage sludge based on quantities of use and fugacity modelling. *Water Sci. Technol.* 46, 105-113.

Khan, SJ and Ongerth, JE (2004) Modelling of prescription pharmaceutical residues in Australian sewage based on quantities of use and fugacity calculations. *Chemosphere* 54, 355-367.

Khan, SJ, Wintgens, T, Sherman, P, Zaricky, J, and Schäfer, AI (2004) Removal of hormones and pharmaceuticals in the Advanced Water Recycling Demonstration Plant in Queensland, Australia. *Water Sci Technol* 50, 15-22.

Khan, SJ and Ongerth, JE (2005) Occurrence and removal of pharmaceuticals at an Australian sewage treatment plant. *Water* 32, 80-85.

Khetan, SK and Collins, TJ (2007) Human Pharmaceuticals in the Aquatic Environment: A Challenge to Green Chemistry. *Chem. Rev.* 107, 2319-2364.

Kolpin, DW, Furlong, ET, Meyer, MT, Thurman, ME, Zaugg, SD, Barber, LB, and Buxton, HT (2002) Pharmaceuticals, hormones and other organic wastewater contaminants in U.S. streams, 1999-2000: a national reconnaissance. *Environ. Sci. Technol.* 36,1202-1211.

Kummerer, K (2004a) Pharmaceuticals in the environment- Scope of the book and introduction. In: Pharmaceuticals in the environment. Second edition. Kummerer, K, Ed. Springer-Verlag, Berlin, 3-11.

Kummerer, K (2004b) Resistance in the environment. *J Antimicro Chemo* 54, 311-320.

Lange, R and Dietrich, D (2002) Environmental risk assessment of pharmaceutical drug substances--conceptual considerations. *Toxicol. Lett.* 131, 97.

Leusch, FDL, Khan, SJ, and Chapman, H (2007) Chemical contaminants in water: Can we measure everything? In: Khan SJ, Stuetz RM, Anderson JM, eds. Water Reuse and Recycling. Sydney, NSW, UNSW Press.

Loos, R, Wollgast, J, Huber, T, and Hanke, G (2007) Polar herbicides, pharmaceutical products, perfluorooctanesulfonate (PFOS), perfluorooctanoate (PFOA), and nonylphenol and its carboxylates and ethoxylates in surface and tap waters around Lake Maggiore in Northern Italy. *Anal. Bioanal. Chem.* 387, 1469-1478.

McDowell, DC, Huber, MM, Wagner, M, Von Gunten, U, and Ternes, TA (2005) Ozonation of carbamazepine in drinking water: Identification and kinetic study of major oxidation products. *Environ. Sci. Technol.* 39. 8014-8022.

Metcalfe, CD, Koenig, BG, Bennie, DT, Servos, M, Ternes, TA, and Hirsch R (2003) Occurrence of neutral and acidic drugs in the effluents of Canadian sewage treatment plants. *Environ. Toxicol. Chem.* 22, 2872-2880.

O'Brien, E, Dietrich, DR (2004) Hindsight rather than foresight: reality versus the EU draft guideline on pharmaceuticals in the environment. *Trends Biotechnol* 22, 326-330.

Pomati, F, Neilan, BA, Manarolla, G, Suzuki, T, and Rossetti, C (2003) Enhancement of intracellular saxitoxin accumulation by lidocaine hydrochloride in the cyanobacterium Cylindrospermopsis raciborskii T3 (Nostocales). *J. Phycol.* 39, 535-542.

Pomati, F, Netting, AG, D. Calamari, D, and Neilan, BA (2004) Effects of erythromycin, tetracycline and ibuprofen on the growth of *Synechocystis* sp. and *Lemna minor*. *Aquatic Toxicol.* 67, 387-396.

Pomati, F, Castiglioni, S, Zuccato, E, Fanelli, R, Vigetti, D, Rossetti, C, and Calamari D, (2006a) Effects of Environmental Contamination by Therapeutic Drugs on Human Embryonic Cells. *Environ. Sci. Technol.* 40, 2442-2447.

Pomati, F, Kellmann, R, Cavaliere, R, Burns, BP, and Neilan, BA (2006b) Comparative Gene Expression of PSP-Toxin Producing and Non-Toxic Anabaena circinalis Strains. *Environ. Int.* 32, 743-748.

Pomati, F, Cotsapas, C, Castiglioni, S, Zuccato, E, and Calamari, D (2007) Gene Expression Profiles in Zebrafish (Danio rerio) Liver Cells Exposed to a Mixture of Pharmaceuticals at Environmentally Relevant Concentrations. *Chemosphere* 70(1):65-73.

Rabiet, M, Togola, A, Brissaud, F, Seidel, JL, Budzinski, H, and Elbaz-Poulichet, F (2006) Consequences of treated water recycling as regards pharmaceuticals and drugs in surface and ground waters of a medium-sized Mediterranean catchment. *Environ. Sci. Technol.* 40, 5282-5288.

Richardson, ML and Bowron, JM (1985) The fate of pharmaceuticals in the aquatic environment. *J. Pharm. Pharmacol.* 37, 1-12.

Schwab, BW, Hayes, EP, Fiori, JM, Mastrocco, FJ, Roden, NM, Cragin, D, Meyerhoff, RD, D'Aco, VJ, and Anderson, PD (2005) Human pharmaceuticals in US surface waters: A human health risk assessment. *Regul. Toxicol. Pharmacol.* 42, 296.

Schwarzenbach, RP, Escher, BI, Fenner, K, Hofstetter, TB, Johnson, CA, von Gunten, U, Wehrli, B (2006) The challenge of micropollutants in aquatic systems. *Science* 313, 1072-1077.

Shon, HK, Vigneswaran, S, and Snyder, SA (2006) Effluent organic matter (EfOM) in wastewater: Constituents, effects, and treatment. *Crit. Rev. Env. Sci. Tech.* 36, 327-374.

Stackelberg, PE, Furlong, ET, Meyer, MT, Zaugg, SD, Henderson, AK, and Reissman, DB (2004) Persistence of pharmaceutical compounds and other organic wastewater contaminants in a conventional drinking-watertreatment plant. *Sci. Total. Environ.* 329:99-113.

Stumpf, M, Ternes, T, Wilken, RD, Rodrigues, SV, and Baumann, W (1999) Polar drug residues in sewage and natural waters in the state of Rio de Janeiro, Brazil. *Sci. Total Environ.* 225, 135-141.

Suzuki, T, Nakazato , K, Shapiro, S, Pomati, F, and Neilan, BA (2004) Effects of synthetic local anaesthetics on the growth of the cyanobacterium *Synechococcus leopoliensis*. *J. Appl. Phycol.* 16, 145-152.

Ternes, TA (1998) Occurrence of drugs in German sewage treatment plants and rivers. *Wat. Res.* 32, 3245-3260.

Ternes, TA, Meisenheimer, M, McDowell, D, Sacher, F, Brauch, HJ, and Haist-Gulde, B (2002) Removal of pharmaceuticals during drinking water treatment. *Environ. Sci. Technol.* 36, 3855-3863.

Ternes, TA, Stuber, J, Herrmann, N, McDowell, D, Ried, A, and Kampmann, M (2003) Ozonation: a tool for removal of pharmaceuticals, contrast media and musk fragrances from wastewater? *Water Res.* 37, 1976-1982.

Ternes, TA and Joss A, Editors In: Human Pharmaceuticals, Hormones and Fragrances - The Challenge of Micropollutants in Urban Water Management. IWA Publishing, London, UK, 2006.

Van de Bogaard, A and Stobberingh, E (1999) Antibiotic usage in animals. *Drugs* 58, 589-607.

von Gunten, U (2003) Ozonation of drinking water: Part I. Oxidation kinetics and product formation. *Water Res.* 37, 1443-1467.

Weber, S, Khan, SJ, and Hollender, J (2006) Human risk assessment of organic contaminants in reclaimed wastewater used for irrigation. *Desalination* 187, 53-64.

Westerhoff, P, Yoon, Y, Snyder, S, and Wert, E (2005) Fate of endocrine-disruptor, pharmaceutical, and personal care product chemicals during simulated drinking water treatment processes. *Environ. Sci. Technol.* 39, 6649-6663.

Yang, S, Carlson, KH (2004) Solid-phase extraction-high-performance liquid chromatography-ion trap mass spectrometry for analysis of trace concentrations of macrolide antibiotics in natural and waste water matrices. *J. Chromatogr. A* 1038, 141-155.

Zuccato, E, Calamari, D, Natangelo, M, and Fanelli, R (2000) Presence of therapeutic drugs in the environment. *Lancet* 355,1789-1790.

Zuccato, E, Castiglioni, S, Fanelli, R, Bagnati, R, and Calamari D (2004) Pharmaceuticals in the environment: changes in presence and concentrations of pharmaceuticals for human use in Italy. In: Pharmaceuticals in the environment, Second edition, Kummerer, K, Ed. Springer-Verlag, Berlin, 45-53.

Zuccato, E, Castiglioni, S, and Fanelli, R (2005a) Identification of the pharmaceuticals for human use contaminating the Italian aquatic environment. *J Hazard Mat* 122, 205-209.

Zuccato, E, Chiabrando, C, Castiglioni, S, Calamari, D, Bagnati, R, Schiarea, S, and Fanelli, R (2005b) Cocaine in surface waters: a new evidence-based tool to monitor community drug abuse. *Environ Health: A Global Access Science Source* 4, 14.

Zuccato, E, Castiglioni, S, Fanelli, R, Reitano, G, Bagnati, R, Chiabrando, C, Pomati, F, Rossetti, C, and Calamari D (2006) Pharmaceuticals in the environment in Italy: causes, occurrence, effects and control. *Environ. Sci. Pollut. Res. Int.* 13, 15-21.

Zuehlke, S, Duennbier, U, Heberer, T (2004) Determination of polar drug residues in sewage and surface water applying liquid chromatography-tandem mass spectrometry. *Anal. Chem.* 76, 6548-6554.

Zwiener, C and Frimmel, FH (2000) Oxidative treatment of pharmaceuticals in water. *Water Res.* 34, 1881-1885.

In: Aquatic Toxicology Research Focus
Editor: Elias P. Svensson, pp. 51-78
ISBN: 978-1-60456-192-0

Chapter 3

MICROBIAL AND CHEMICAL ASSAYS TO DETERMINE THE ORIGIN OF FAECAL POLLUTION AND THE PRESENCE OF WATERBORNE PATHOGENS

***M.M. Ehlers*[1], *A. Sundram*[2], *M.M. Kock*[1] *and N. Potgieter*[3]**

[1]Department of Medical Microbiology,
University of Pretoria/National Health Laboratory Services, South Africa
[2]Umgeni Water, Kwazula Natal, South Africa
[3]Department of Microbiology, University of Venda, South Africa

ABSTRACT

In the last decade, the emphasis has shifted to the physical and microbiological quality of water sources as reports have shown that there are an increasing percentage of surface- and groundwater that are being contaminated with waterborne pathogens. Increased land use, rural development, informal settlements and inadequate infrastructure for water and sanitation have increased the risk of faecal contamination of these water sources. This is compounded by alterations in the hydrological cycle, associated with global climate change that increases the magnitude and frequency of run-off events. Waterborne diseases and epidemics result from the ingestion of faecally polluted water containing enteric bacterial pathogens, human-specific enteric viruses and protozoan parasites. Enteric bacteria have traditionally been used as indicators of faecal contamination of source, drinking and recreational water. However, research has demonstrated that these bacteria may not be appropriate indicators of pathogenic viruses and the need for reliable index or indicator organisms of these pathogens have been widely recognized. Studies conducted over the past couple of years have investigated, evaluated and compared bacteriophage and chemical assays as indicators of waterborne pathogens. In one study, the potential of F-RNA phage genotyping and chemical analysis of faecal sterols were evaluated to indicate if faecal pollution of water sources were of human or animal origin. When applied to environmental samples, F-RNA phage genotyping provided useful information in certain instances and confirmed the presence of either human or animal faecal pollution. Chemical assays yielded more conclusive results when F-RNA phage genotyping results obtained from river water samples were ambiguous. In another study F-RNA phage genotyping was used to determine the origin of faecal pollution in traditional and CDC water containers in rural settlements. Results showed no difference between the different containers; both human and animal phage subtypes were detected. Although the detection of each specific pathogen seems to be the ideal, this is not always possible due to expensive analysis and the need of well-equipped laboratories and skilled personnel. It can be concluded that bacterial and bacteriophage indicators are still the best alternative and most cost-effective tools currently available and will continue to play a major role in the determination of the microbial quality of water.

INTRODUCTION

Diarrhoeal diseases are responsible for approximately 2.5 million deaths annually in developing countries, affecting children younger than five years, especially those in areas devoid of access to potable water supply and sanitation (Kosek *et al.*, 2003; Obi *et al.*, 2003; Lin *et al.*, 2004; Obi *et al.*, 2004). One of the Millennium Declaration goals was to reduce the amount of people without safe drinking water by half in the year 2015 (UN, 2000). Safe drinking water for human consumption should be free from pathogens such as bacteria, viruses and protozoan parasites; meet the standard guidelines for taste, odour, appearance and chemical concentrations and must be available in adequate quantities for domestic purposes (Kirkwood, 1998). However, inadequate sanitation and persistent faecal contamination of water sources is responsible for a large percentage of people in both developed and developing countries not having access to microbiologically safe drinking water and suffering from diarrhoeal diseases (WHO, 2002a; WHO, 2002b).

Political upheaval, high numbers of refugees in some developing countries and the global appearances of squatter camps and shanty rural towns, which lack proper sanitation and water connections have contributed to conditions under which disease causing microorganisms can replicate and thrive (Leclerc *et al.*, 2002; Sobsey, 2002; Theron and Cloete, 2002). The people most susceptible to waterborne diseases include young children, the elderly, people suffering from malnutrition, pregnant woman, immunocompromised individuals, people suffering from chemical dependencies and persons predisposed to other illnesses like diabetes (Sobsey *et al.*, 1993; Gerba *et al.*, 1996; Grabow, 1996; Leclerc *et al.*, 2002; Theron and Cloete, 2002). Furthermore, an increasing number of people are becoming susceptible to infections with specific pathogens due to the indiscriminate use of antimicrobial drugs, which have lead to the selection of antibiotic resistant bacteria and drug resistant protozoa (WHO, 2002c; NRC, 2004).

In developing countries, many people are living in rural communities and have to collect their drinking water some distances away from the household and transport it back in various types of containers (Sobsey, 2002). Microbiological contamination of the water may occur between the collection point and the point-of-use in the household due to unhygienic practices causing the water to become a health risk (Sobsey, 2002; Gundry *et al.*, 2004; Moyo *et al.*, 2004).

Waterborne diseases and epidemics have been associated with water supplies polluted by faecal matter containing a range of human-specific bacterial, viral and protozoan pathogens (Jagals *et al.*, 1995; Sinton *et al.*, 1998; Scott *et al.*, 2002). Research has indicated that more than 100 types of pathogenic human viruses may be present in faecally polluted water

(Havelaar *et al.*, 1993; Scott *et al.*, 2002). It must be noted that some waterborne pathogens such as *Cryptosporidium* spp. and a limited number of bacteria and viruses may be of animal origin (Sinton *et al.*, 1998). Information on the human or animal origin of faecal pollution is therefore important because it gives an indication of the possible risk of infection and the treatment that might be needed to control the transmission of waterborne diseases (Jagals *et al.*, 1995).

The major causes of faecal contamination of water sources such as rivers, dams, groundwater as well as drinking water are: i) release of partially treated sewage or sewage leakage, ii) leaching of poorly maintained septic tanks, iii) improper management of farm waste and iv) run-off of faecal matter during rainy periods (Isobe *et al.*, 2004). The management of faecal contamination of water can only improve if the sources of pollution are identified. Water analysis and detection tools to distinguish between human and animal faecal sources are therefore needed (Schaper *et al.*, 2002a).

Waterborne enteric viruses can reach large human populations and cause diseases such as gastroenteritis and hepatitis (Beekwilder *et al.*, 1996). Human enteric viral diseases are considered to be predominantly associated with the ingestion of human-derived wastes because of the host-specific nature of enteric viruses (Calci *et al.*, 1998; Sinton *et al.*, 1998). Studies conducted by Havelaar and colleagues (1993) showed that viruses are more persistent than bacteria in water environments and are not readily removed by water treatment processes. Water treatment and catchment management strategies based on bacteriological indicators seems inadequate tools to be used to assess the virological safety of water (Havelaar *et al.*, 1993; Cole *et al.*, 2003; Skraber *et al.*, 2004).

Bacteriophages have the closest correlation with enteric viruses making them invaluable indicators of the possible presence and behaviour of enteric viruses (Geldenhys and Pretorius, 1989; Brion *et al.*, 2002; Skraber *et al.*, 2004). Although enteric viruses are the ultimate viral indicator, analyses for coliphages (phages) such as somatic and male specific F-RNA phages are easier and more rapid to enumerate and less expensive compared to the detection of specific enteric viruses (Rhodes and Kator, 1991). The International Association for Water Pollution Research and Control (IAWPRC) (1991) as well as Donnison and Ross (1995) reported that phages have similar behaviour patterns to enteric viruses in the environment. Phages are found in abundance in wastewater, since they are excreted in human and animal faeces (Armon, 1993). Depending on the conditions prevailing, it was found that phages outnumbered enteric viruses by a factor of at least a 1 000 or more (Grabow *et al.*, 1984). In addition, phages are more resistant than enteric viruses to water treatment and disinfection processes as well as to unfavourable environmental conditions shown by studies conducted at pH 4 and pH 10 and temperatures ranging from 4°C to 37°C (Grabow *et al.*, 1984; Schaper *et al.*, 2002a; Allwood *et al.*, 2003; Cole *et al.*, 2003).

The male specific F-RNA phages have been associated exclusively with animal or human faecal contamination and have been classified into four groups (also referred to as serological groups or subgroups) (Havelaar *et al.*, 1990). Two groups have been reported to originate from animal sources and the other two groups originating from human sources (Osawa *et al.*, 1981; Furuse, 1987; Havelaar *et al.*, 1990; Hsu *et al.*, 1995; Schaper and Jofre, 2000; Schaper *et al.*, 2002b). Researchers have successfully developed and exploited genotyping techniques to group F-RNA phage isolates into these four respective groups (Hsu *et al.*, 1995; Beekwilder *et al.*, 1996; Griffin *et al.*, 2000).

In addition to F-RNA phage typing to distinguish between faecal pollution of humans and animals, the chemical analysis of faecal sterols such as cholesterol and coprostanol have yielded valuable results in attempts to distinguish between faecal pollution of human and animal origin (Nguyen *et al.*, 1995; Leeming and Nichols, 1996). The profiles of faecal sterols such as cholesterol and coprostanol differ in different warm-blooded animals and are gaining wide use as biomarkers indicating faecal pollution (Nguyen *et* al., 1995; Leeming *et al.*, 1996). The addition of a silylating agent converts these compounds into trimethylsilyl (TMS) ethers, which are more stable for gas chromatographic assays (Elhmmali *et al.*, 2000). The coupling of a mass spectrometer to a gas chromatograph proved suitable for the separation and identification of the spectrum of compounds belonging to faecal sterols and TMS ethers on the basis of, which meaningful data have been reported to distinguish between human and animal excreta (Pool, 1978; Nguyen *et al.*, 1995; Leeming and Nichols, 1996). The aim of this first study was to use genotyping of F-RNA phages and chemical assays for the faecal sterols, cholesterol and coprostanol, to track faecal pollution to human or animal sources in different wastewater and environmental samples.

Poor sanitation, unhygienic practices and close living associations between people and animals in rural communities increase the risk of zoonoses and add to faecal contamination of stored drinking water. Point-of-use interventions can improve the microbiological quality of household drinking water and a combination of microbial and chemical indicator tests could identify the origin of faecal pollution. Following the first study, a second study was conducted to evaluate the improvement of the microbiological quality of drinking water in rural households. The implementation of intervention strategies, which included the use of traditional storage containers as well as an improved safe storage container (CDC, USA), with or without the addition of a sodium hypochlorite solution as well as the origin of faecal pollution were investigated. However, for the purpose of this chapter we only focused on the origin of faecal contamination in the water sources and household stored water using male specific F-RNA group genotyping.

Methodology for the Identification of the Origin of Faecal Pollution in Water Sources

Genotyping of F-RNA Phages

The F-RNA phages have been classified into 4 groups based on their phylogenetic diversity, eg. MS-2 in group 1, GA in group 2, Qβ in group 3 and SP in group 4 (Havelaar and Hogeboom, 1984; Furuse, 1987). Serotyping of F-RNA phages has been suggested as a method to determine the origin of faecal pollution (Schaper and Jofre, 2000). However, serotyping was found to be laborious and required antiserum reagents that were scarce and not readily available (Havelaar *et al.*, 1990; Hsu *et al.*, 1995). Genotyping was suggested as the better alternative to serotyping because several F-RNA phages were reported as serological intermediates between group 1 and 2 and between groups 3 and 4, possibly resulting in ambiguous serotyping results (Hsu *et al.*, 1995). Hsu and co-workers (1995) researched highly specific nucleic acid probes for the four groups of F-RNA phages that were successfully used for genotyping F-RNA phages. The presence of F-RNA phages in a water sample generally indicates pollution by wastewater, which contains human and/or animal faeces (DWAF, 1996). The F-RNA phage serogroups 1 and 4 occurred in wastewater from animal sources while serogroups 2 and 3 were found in wastewater from human sources such as hospital wastewater (Osawa *et al.*, 1981; Havelaar *et al.*, 1986; Havelaar *et al.*, 1990; Hsu *et al.*, 1995; Schaper *et al.*, 2002b).

Schaper and Jofre (2000) confirmed that groups 2 and 3 predominated in municipal sewage illustrating that these genotypes were of human origin. Research conducted by Schaper *et al.* (2002b) found that the association of groups 2 and 3 to human excreta and groups 1 and 4 to animal excreta significant. Other research conducted by Griffin *et al.* (2000) found that genotyping of F-RNA phages could be used successfully to help distinguish between faecal pollution of animal or human origin.

It is noteworthy that all isolates from animal faecal sources were grouped as groups 1 and 4 with the exception of isolates from pig and piglet faeces, which often contained groups 2 or 3 F-RNA phages (Hsu *et al.*, 1995; Cole *et al.*, 2003). This suggests that group classification will not always distinguish between human and porcine faecal contamination (Osawa *et al.*, 1981; Furuse, 1987; Havelaar *et al.*, 1990; Hsu *et al.*, 1995). This occurrence could be explained by the dietary and living conditions of pigs that have historically involved exposure to human faecal wastes and thus human-specific F-RNA phages (Hsu *et al.*, 1995). Once introduced into pig populations, these F-RNA phages could persist because the gastrointestinal physiology and flora of pigs are similar to those of humans (Hsu *et al.*, 1995).

Gene Probes Used for F-RNA Phage Genotyping

Gene probes are homologous nucleic acid material that could bind with or hybridise to specific target nucleic acids and facilitate the detection of specific target nucleic acids (Hurst *et al.*, 1989). The development of F-RNA phages probes involved searching for stretches of 20 to 30 nucleotides which showed complete conservation within one subgroup, while being absent from other subgroups (Table 1) (Hsu *et al.*, 1995; Beekwilder *et al.*, 1996).

Table 1. Nucleotide sequences of oligomers used for probing F-RNA phages (Beekwilder *et al.*, 1996)

Group	Probe Sequence	Locus
1	CGCATCTAAGGTATGGACCATCGAGA	Maturation protein
2	AATCGTTCAGGAAGTGAGATTCAAAACC	Maturation protein
3	AGCCAGAGATTACCAGCAGTAGC	5' Non-translated region
4	GGCATAGATTCTCCTCTGTAGTGCG	5' Non-translated region

Various computer software gene analysis and alignment programmes were used to select oligonucleotide sequences having the least sequence homology with phages of heterologous groups (Hsu *et al.*, 1995; Beekwilder *et al.*, 1996). These oligoprobes, which were end-labeled with digoxigenin, were applied in DNA-RNA hybridisations and hybrids were observed for colorimetric or chemiluminescent detection (Hsu *et al.*, 1995; Beekwilder *et al.*, 1996; Griffin *et al.*, 2000).

Gene Hybridization of F-RNA Isolates

Nucleic acid hybridisation involves the use of nucleic acid probes for detecting viral nucleic acid that has been trapped on filter material or membranes (Hurst *et al.*, 1989). The method for transfer of phages from zones of lysis (plaques) on a host cell lawn to candidate membrane filters and fixation on the membranes were optimised to yield the best possible detection following hybridisation (Hsu *et al.*, 1995; Beekwilder *et al.*, 1996; Griffin *et al.*, 2000). Many types of filters including nitrocellulose, non-charged and charged nylon Hybond N filters were tested for their ability to retain and expose phage RNA from plaques (Hsu *et al.*, 1995; Beekwilder *et al.*, 1996). The Hybond N+ nylon membranes were most satisfactory and reproducibly yielded sharp plaque signals, contrasting clearly to a low background, even when plaques had a small diameter (Hsu *et al.*, 1995; Beekwilder *et al.*, 1996; Griffin *et al.*, 2000; Schaper and Jofre, 2000).

Following fixation of phages onto membranes, the phage RNA would have to be released and denatured, thereby exposing its bases to the complementary oligonucleotides (Beekwilder *et al.*, 1996; Schaper and Jofre, 2000). This was achieved by alkaline treatment with NaOH followed by neutralisation with sodium acetate (Hsu *et al.*, 1995; Beekwilder *et al.*, 1996; Griffin *et al.*, 2000; Schaper and Jofre, 2000). Although alkaline treatment could lead to hydrolysis of the RNA chain, studies have found that 1 minute exposure to 0.05 M NaOH leaves the RNA sufficiently intact to serve as a target for the oligonucleotide probes (Beekwilder *et al.*, 1996; Uys, 1999).

Following fixation and denaturation, the membranes need to undergo a pre-hybridisation step in preparation for the binding of the probes (Hsu *et al.*, 1995). Research done by Uys (1999) found that a modified pre-hybridisation step used by Beekwilder *et al.* (1996) was more efficient than the method used by Hsu *et al.* (1995). This modification included increasing the pre-hybridisation time to 1 hour, which resulted in the least background colour development when chemiluminescent detection was performed (Uys, 1999).

Various studies used probe concentrations ranging from 2.5 – 5 $pmol.ml^{-1}$ with hybridisation conditions ranging from 37°C – 45°C for 1 hour to overnight (Hsu *et al.*, 1995; Beekwilder *et al.*, 1996; Griffin *et al.*, 2000; Schaper and Jofre, 2000). Studies conducted by Uys (1999) found that overnight incubation of the viral nucleic acids with 2.5 $pmol.ml^{-1}$ of labeled probe at 37°C allowed maximum probe-target interaction. This lower probe concentration that was used, yielded better results when chemiluminescent detection was performed (Uys, 1999).

Detection of DNA:RNA Hybrids of F-RNA Phages

Colorimetric and radioactive detection techniques used by Hsu and co-workers (1995) and Beekwilder and colleagues (1996) have been superceded by chemiluminescent detection techniques used by researchers such as Griffin and colleagues (2000) as well as Schaper and Jofre (2000). The colorimetric detection technique with the DIG Nucleic Acid Detection Kit showed hybrids to appear as dark-blue circular spots on membranes (Hsu *et al.*, 1995; Uys, 1999). Research conducted by Uys (1999) found that colorimetric detection was time consuming and some hybrids tended to appear as light to dark blue spots resulting in subjective assigning of probes to subgroups. Chemiluminescent detection is performed with

Lumigen-CSPD and hybridised plaques appeared as dark-black circular spots on the X-Ray film (Schaper and Jofre, 2000).

Faecal Sterols Used as a Tool to Trace the Origins of Faecal Pollution in Water

There are at least seventeen sterols found in the faeces of humans and animals including sterols such as cholesterol and campesterol and stanols such as coprostanol and cholestanol (Leeming *et al.*, 1996). Leeming and colleagues (1996) discussed that bacterial indicators are unable to distinguish between animal and human pollution whereas the use of biomarkers, in particular faecal sterols such as cholesterol and coprostanol are advantageous as they can distinguish between the origins of faecal pollution. Biomarkers are organic compounds (Writer *et al.*, 1995) that have maintained their structural integrity for their source to be recognised (Leeming *et al.,* 1996). These chemical signatures overcome many of the shortcomings (e.g. die-off, lack of correlation with faecal pathogens) of classical microbiological indicators of sewage pollution (Nichols *et al.*, 1996; Elhmmali *et al.*, 2000). Pollution from animal and human wastes is a cause of deteriorating water quality and the management of the problem could depend on tracing the source of pollution using the sterol profiles of human and animal faeces (Leeming *et al.,* 1996; Sinton *et al.*, 1998).

The sterol profiles from various animal faeces showed that there are differences in both the content and composition of faecal sterols between warm-blooded animals (Leeming *et al.*, 1996; Leeming *et al.*, 1998). The source-specificity of faecal sterols is caused by a combination of three main factors (Leeming *et al.*, 1996). The first is the animal's diet that has a different sterol profile so the proportions of sterol precursors entering the digestive tract are different (Leeming *et al.*, 1996; Nichols *et al.*, 1996). Secondly, even with low dietary intake of sterols, endogenous sterols are biosynthesised by mammalian animals and discharged to the digestive tract for example, humans on low cholesterol diets will still synthesise cholesterol (Leeming *et al.*, 1996). Thirdly, and most importantly, anaerobic bacteria in the digestive tract of some animals biohydrogenate sterols to stanols of various isomeric configurations (Leeming *et al.*, 1996). The combination of these three factors determines "the sterol fingerprint" of the faeces for any animal (Nichols *et al.*, 1996).

Studies conducted by Nichols *et al.* (1996) found that the total sterol content in human faeces was 3 000 – 8 800 $\mu g.g^{-1}$, being five times higher than the concentrations detected from other animal samples. Human faeces contained as much as ten times more coprostanol on a dry weight basis than faeces analysed from animals (Leeming *et al.*, 1996; Scott *et al.*, 2002). The sterol profile of pigs was found to be similar to that of humans, although the absolute concentrations were about 10 fold lower (Leeming *et al.*, 1996; Nichols *et al.*, 1996; Scott *et al.*, 2002). Murtuagh and Bunch (1967) found that the concentration of coprostanol excreted by humans was at least seventeen times higher than that excreted in pigs and cows. Later research found that coprostanol concentrations in human faeces were one to three orders of magnitude greater than that found in other animals (Nichols *et al.*, 1996).

Research conducted by Leeming and colleagues (1996) indicated that the ratios of coprostanol:cholesterol in humans were as high as 10:1. This was opposed to certain animals that had relatively similar coprostanol:cholesterol ratios and others that had cholesterol concentrations that were more than two orders of magnitude higher than coprostanol concentrations (Leeming *et al.*, 1996). Nguyen and co-workers (1995) cited that these two major sterols: cholesterol and coprostanol are poorly soluble in water and have been detected in wastewater in concentrations exceeding 1 000 $ng.L^{-1}$ (Leeming and Nichols, 1996). Assays for the sterols coprostanol and cholesterol could prove to be better than traditional microbiological techniques for detecting human sewage pollution versus faecal contamination from animal sources (Leeming and Nichols, 1996).

Cholesterol Used as a Faecal Indicator

Cholesterol originates from the diet, liver, intestinal epithelium and other tissues in the body (Macdonald *et al.*, 1983). Cholesterol occurs widely in the human body, but not all of the biological functions of cholesterol are known (Graham Solomons, 1980). This widely occurring steroid is known to serve as an intermediate in the biosynthesis of steroid hormones and of bile acids (Graham Solomons, 1980). However, it is reported that more cholesterol is present in the human body than is needed for the biosynthesis of steroid hormones (Graham Solomons, 1980). As cholesterol is also found in eggs, milk, lard and wool grease, it is not as specific an indicator of faecal pollution as coprostanol (Dutka *et al.*, 1974; Dutka and El-Shaarawi, 1975).

Coprostanol Used as a Faecal Indicator

According to Martin and co-workers (1973), coprostanol in human faeces resulted from the biohydrogenation of cholesterol by the intestinal microflora. The biohydrogenation of the unsaturated sterol cholesterol to coprostanol in the small intestine of mammalian animals is carried out by a mixed population of faecal bacteria such as *Bifidobacterium* spp., *Clostridium* spp., *Bacteroides* spp. and *Eubacterium* spp. (Dutka *et al.*, 1974; Macdonald *et al.*, 1983; Venkatesan and Santiago, 1989; Vivian, 1986; Nichols *et al.*, 1996; Elhmmali *et al.*, 2000). In the metabolic formation of coprostanol, the double bonds between carbon atoms 5 and 6 in the B ring of cholesterol are enzymatically hydrogenated resulting in the formation of three stereo-isomers; coprostanol (5β-coprostanol), epicoprostanol (5β-epicoprostanol) and dihydrocholesterol (5α-cholestanol) (Matusik *et al.*, 1988).

Coprostanol is a major sterol found in human faeces (24% - 89% of total steroids) and has been proven to be a promising indicator of faecal pollution by various studies of coprostanol concentrations in the water and sediments of rivers, lakes and estuaries (Macdonald *et al.*, 1983; Scott *et al.*, 2002; Isobe *et al.*, 2004). The intestinal tract of adults contains approximately 1 kg of bacteria (10^{14} organisms), the bulk of which is in the colon and both dietary and endogenous substances are destined to come into intimate contact with this population of more than 400 species in the lower gut (Macdonald *et al.*, 1983). As a result, human faeces and municipal wastewater sludge contain a high concentration of coprostanol (Chan *et al.*, 1998).

Coprostanol is aerobically degraded by natural bacteria and thus the presence of coprostanol in an aerobic environment would indicate recent sewage contamination (Vivian, 1986). The reported half-life of coprostanol in aerobic conditions are generally <10 days at

20°C, thus the presence of coprostanol in an aerobic environment can be considered to be of recent faecal input to the water (Isobe *et al.*, 2004). Coprostanol has been suggested to be useful in tracing sewage effluents sterilised by chlorination since it is unaffected by the process unlike faecal bacteria (McCalley *et al.*, 1981; Vivian, 1986). Another attribute that makes coprostanol a good chemical indicator of human pollution is that it is a primary sterol in domestic wastes and it is unaffected by physical factors like temperature and salinity (Brown and Wade, 1984; Writer *et al.*, 1995). However, humans may vary as to the amount of coprostanol they excrete (Leeming and Nichols, 1996).

Faecal Sterol Formation and Structure

The structural differences between cholesterol and the three common saturated isomeric (stanol) products are shown in Figure 1. The seemingly small changes in orientation of the hydrogen atom at the carbon-5 position and the hydroxyl group at the carbon-3 position result in different compounds which can be used as biomarkers (Leeming *et al.*, 1996; Nichols *et al.*, 1996). The 5α configuration of the C_{27} stanol, 5α-cholestanol, is the thermodynamically stable isomer (Figure 2.3), whereas 5β-coprostanol and 5β-epicoprostanol are produced only through the mediation of anaerobic bacteria (Leeming *et al.*, 1996; Nichols *et al.*, 1996).

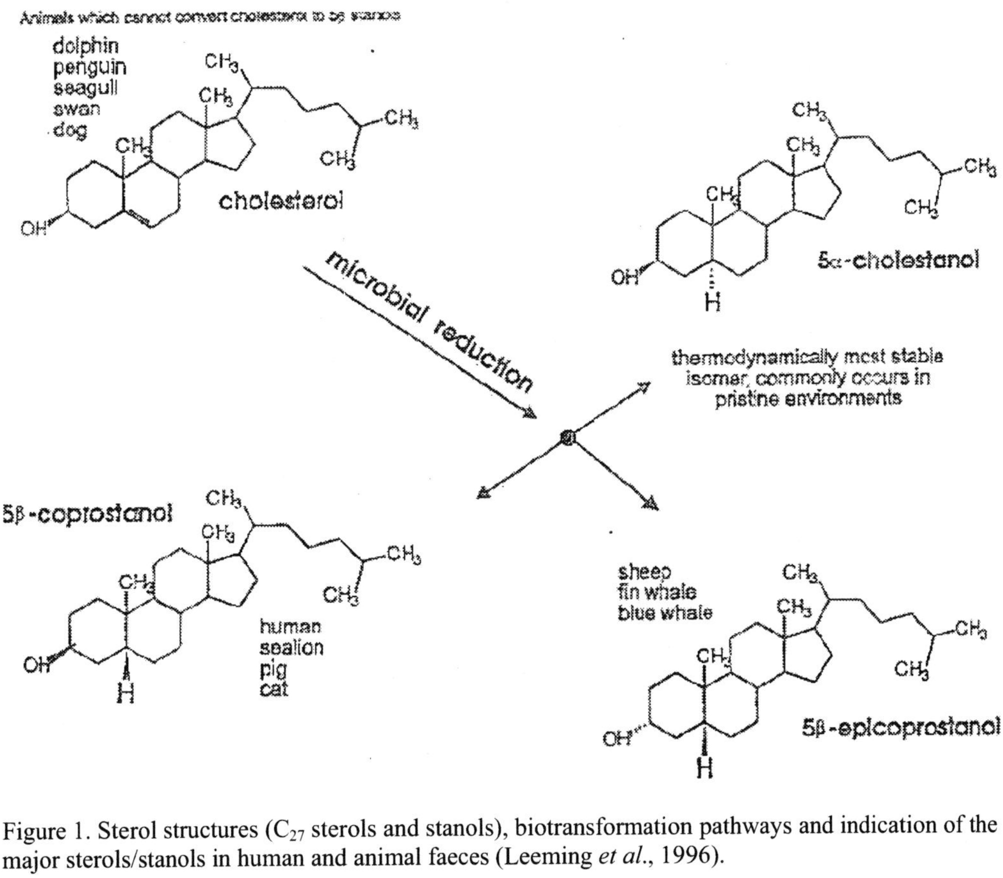

Figure 1. Sterol structures (C_{27} sterols and stanols), biotransformation pathways and indication of the major sterols/stanols in human and animal faeces (Leeming *et al.*, 1996).

Anaerobic bacteria preferentially biohydrogenate Δ^5-sterols to their corresponding 5β(H), 3β(OH)-stanols (Leeming *et al.*, 1996). Matusik and co-workers (1988) reported that coprostanol, cholesterol and 5α-cholestanol are the principal faecal sterols found in human faeces. Anaerobic bacteria in the guts of other animals either preferentially biohydrogenate Δ^5-sterols to one or more corresponding stanols (Figure 1), or lack the ability to biohydrogenate Δ^5-sterols at all (Leeming *et al.*, 1996).

Extraction of Faecal Sterols From Water

The ideal solvent or solvent mixture for extracting lipids should be sufficiently polar to remove all lipids from the mixture but should not chemically react with the lipid itself (Christie, 1982). Lipids, which contain no markedly polar groups, for example cholesterol, are highly soluble in hydrocarbon solvents such as hexane, benzene or cyclohexane as well as in slightly more polar solvents like chloroform but are rather insoluble in polar solvents such as methanol (Bligh and Dyer, 1959; Christie, 1982). The solubility of faecal sterols in alcoholic solvents increases with the chain length of the hydrocarbon moiety of the alcohol so these compounds become more soluble in ethanol and completely soluble in n-butanol (Christie, 1982).

During the formation of silyl ethers, the character of the sterol molecule is changed from polar and active to non-polar and inert, which is generally coupled with an increase in volatility and thermal stability, which are the properties most desirable for gas chromatography (Pool, 1978). A powerful silyl donor such as bis(trimethylsilyl) trifluoroacetamide (BSTFA) is widely used to form silyl derivatives of organic compounds such as faecal sterols (Pool, 1978; Writer *et al.*, 1995; Leeming *et al.*, 1996; Nichols *et al.*, 1996; Elhmmali *et al.*, 2000). Thus the addition of the silylating agent converts these organic compounds into TMS ethers (Elhmmali *et al.*, 2000). The coupling of a mass spectrometer to a gas chromatograph has proved to be a good method of separating and identifying the components of a complex organic mixture in a simple yet highly diagnostic manner (Pool, 1978).

Gas Chromatography-Mass Spectrometry Used for the Detection of Faecal Sterols

Gas chromatography (GC) is the physical separation of two or more compounds based on their differential distribution between a stationary phase and a gaseous mobile phase (Fowles, 1998). Compounds to be analysed must be volatile to be present in the gas phase under experimental conditions so that it can be transported through the fused silica capillary columns where compounds are separated (Fowles, 1998). Gas chromatographs require a supply of carrier gas, such as nitrogen, helium or hydrogen, of sufficient quality and pressure to achieve the desired separations (Fowles, 1998). Therefore, the basic principle of GC is that the greater the affinity of the compound for the stationary phase, the more the compound will be retained by the column and the longer it will take before it is eluted and detected (Fowles,

1998). A detector converts the column eluent to an electrical signal that is measured and recorded (Watson, 1985; Furniss *et al.*, 1989).

Mass Spectrometry (MS) is one kind of GC detector (Watson, 1985). As the separated sample component molecules elute form the column to the inside of the MS they are bombarded with energy causing them to loose an electron and form ions with a positive charge (Watson, 1985). Some of the bonds holding the molecule together are broken in the process, and the resulting fragments may rearrange or break up further to form stable fragments (Watson, 1985; Furniss *et al.*, 1989). Due to the natural laws that govern the relative strengths of chemical bonds, a given compound will ionise, fragment and rearrange reproducibly under a given set of conditions (Watson, 1985). A mass spectrum is a recording of the mass of each of the ionised fragments, representing a unique fingerprint of a molecule that can be used in identification (Watson, 1985; Furniss *et al.*, 1989). Molecular fragments appear in a mass spectrum and provide clues to the identity and molecular structure of the parent molecule similar to the way pieces of a jigsaw puzzle provide clues to the structure of the intact puzzle (Watson, 1985; Furniss *et al.*, 1989). A mass spectrum can be described as a molecular fingerprint consisting of a graph showing the mass:charge ratio (in Daltons or atomic mass units) versus abundance data for ions in the sample molecule and its fragments where the largest peak in the spectrum is called the base peak (Watson, 1985; Furniss *et al.*, 1989). The mass:charge ratio of any fragment therefore is normally equal to the mass of that fragment (Watson, 1985).

Mass spectrometers have been used to detect and quantify compounds at concentrations as low as only a few parts per billion (Watson, 1985). Gas chromatographic data is two dimensional, i.e. a plot of signal versus time (Fowles, 1998). A mass spectrometer adds another dimension to GC data, i.e. abundance and mass versus time (Watson, 1985; Furniss *et al.*, 1989). The GC-MS chromatogram represents the sum of all ions detected (Watson, 1985; Furniss *et al.*, 1989). Quantification with GC-MS can be accomplished by using calibration curves obtained from injecting known amounts of standards and recording the responses of the MS for each amount injected making up these curves (Watson, 1985; Furniss *et al.*, 1989; Fowles, 1998). As the compound is injected at an unknown concentration, the amount of the unknown can be calculated by relating the response to the amount on the curve (Watson, 1985; Furniss *et al.*, 1989).

Sample Collection for F-RNA Genotyping and Chemical Assays (Study 1)

Wastewater samples (500 ml) representative of human faecal pollution were collected from two local hospitals in Pietermaritzburg, Kwa-Zulu Natal, South Africa. Wastewater (500 ml) from a cattle feedlot in the Umgeni River catchment and wastewater from a piggery in the Pietermaritzburg midlands were collected as being representative of cattle and pig faecal pollution. Chicken wash and wastewater (500 ml) from a poultry farm and an abattoir in the Cato Ridge area was collected as representative chicken faecal pollution. River water samples (500 mL) were collected from the Slangspuit, Baynespruit, uMsunduze and Dorpspruit rivers. River water samples analysed for faecal sterols were collected in 2 L pyrex glass bottles.

Sample Collection for F-RNA Genotyping During an Intervention Study (Study 2)

The second study focused on rural communities in the Vhembe region of the Limpopo Province, South Africa and investigated the microbiological quality of drinking water in rural households, evaluated the implementation, compliance and sustainability of intervention strategies such as the CDC safe storage container and chlorine practices, assessed the survival of selected pathogens and investigated sources of faecal contamination in household stored water. However, in this chapter we will just focus on the genotyping of F-RNA phages in the determination of the origin of faecal pollution.

Forty households were randomly selected in two rural villages (Vembe region, Limpopo province, South Africa). Two rounds of water collection from the households and the water sources were carried out over a period of 5 months. Water samples (2 L) were collected aseptically in sterile Nalgene water collection bottles (Merck, Darmstadt, Germany) from 7 communal taps in village 1 and from 4 points on the Sambandou River in village 2.

Two water storage containers in each household were randomly selected and vigorously shaken before water samples were collected. Water samples (1 L each) were collected aseptically in sterile Nalgene water collection bottles (Merck, Darmstadt, Germany) from each of the two selected storage containers in each household on both rounds of water collection. The two water samples from each household were pooled into one sample (2 L) representative of the household container water, placed on ice and transported to the laboratory for further analyses within 8 h. After the first round of water sampling, 20 households in each village were randomly selected and provided with two CDC safe storage containers each. These households were requested to use the CDC safe storage containers instead of the traditional household storage containers. The water samples obtained from the households using the CDC safe storage container were compared with water samples obtained from the households in the same village using traditional storage containers in order to determine the impact of an improved storage container on the origin of faecal pollution.

Water contaminated with human faeces is regarded as a greater risk to human health as there is more likely a chance that it would contain human-specific enteric pathogens (Scott *et al.*, 2002). It has become increasingly important to distinguish faecal pollution between human and animal. Various microbiological and chemical methods have been demonstrated to help identify the source of faecal pollution, thereby assisting in establishing the associated risks (Hsu *et al.*, 1995; Beekwilder *et al.*, 1996; Leeming and Nichols, 1996; Griffin *et al.*, 2000; Schaper *et al.*, 2002a; Isobe *et al.*, 2004).

RESULTS AND DISCUSSION

In the first study the isolation of F-RNA phages, following the standard ISO-10705–1 (1995) double-layer agar technique, was found to be simple and easy once established in the laboratory. Plaques that developed were opaque and not as clear as those formed by typical somatic phages. Due to the nature of some of the environmental samples, a fair amount of background bacterial growth did occur. The identification of F-RNA phages by nucleic acid hybridisation has clear advantages over other techniques since it is genome-based and very specific.

Evaluation of DNA:RNA Hybridisation Techniques Using Control Phages

The hybridisation protocols established by Hsu *et al.* (1995) and Beekwilder *et al.* (1996) and optimised by Uys (1999) were carefully followed. The Hybond N+ nylon membranes were most satisfactory and reproducibly yielded sharp plaque signals, contrasting clearly to a low background, even when plaques had a small diameter (Hsu *et al.*, 1995; Beekwilder *et al.*, 1996; Griffin *et al.*, 2000; Schaper and Jofre, 2000). Overnight incubation of the viral nucleic acids with 2.5 $pmol.ml^{-1}$ of labeled probe at 37°C allowed maximum probe-target interaction and yielded good results when chemiluminescent detection was performed. Initial experiments to fix phage RNA onto the membranes using a standard UV lamp (365 nm), in the absence of a UV Transilluminator, placed over the membranes in a dark box proved to be unsuccessful. This was substituted with baking the membranes in a hot air oven at 80°C for 2 hrs. The DNA:RNA hybrids appeared as dark spots on the X-Ray film. Spots that developed were clearly visible and exposure time of 3.5 – 4 min was adequate in generating enough signal to allow detection on the film. Some background colour development did occur on some films whenever F-RNA phage plates were left in the cold room for longer periods, usually more than a week, allowing bacterial re-growth to occur.

Researchers such as Hsu *et al.* (1995) and Beekwilder *et al.* (1996) developed probes for the four different subgroups of F-RNA phages by searching for stretches of 20 to 30 nucleotides, which showed complete conservation within one subgroup, while being absent from other subgroups. The F-RNA phages GA, SP, MS2 and Qβ representative of each of the four F-RNA phage groups were used as controls for the DNA:RNA hybridisation. These control phages confirmed that there was no cross reactivity between probes of the different F-RNA phage genotypes.

Genotyping of F-RNA Phages from Wastewater and River Samples

Although both the Cabelli E. coli and Havelaar S. typhimurium WG 49 hosts detected other phages such as somatic and F-DNA phages, it was reported that more than 90% – 95% of phages detected from sewage and environmental samples using S. typhimurium WG 49 as the host, were F-RNA phages (Havelaar and Hogeboom, 1984; Debartolomeis and Cabelli, 1991; Hsu et al., 1995; Grabow et al., 1997). With this in mind it would be safe to assume that most of the phages isolated from the environmental samples using S. typhimurium WG 49 as the host were F-RNA phages, confirmed by the positive genotyping results obtained.

Sample sites representative of the predominant animal sources in the catchment such as; cattle, chicken and pigs, were selected from accessible sites at abattoirs, feedlots and piggeries. Wastewater samples obtained from two private hospitals in Pietermaritzburg were selected as representative human faecal input. Samples were collected from these different sites and F-RNA phages were isolated following the standard ISO-10705–1 (1995) methodology. These phages were hybridised with probes specific for each of the F-RNA phage groups and detected using chemiluminescent detection techniques as described earlier.

The results showed that F-RNA groups 3 and 4 were the only two groups detected in pig wastewater. Unlike previous studies conducted by Hsu *et al.* (1995), F-RNA phage group 2 was not detected in the pig wastewater tested. Another interesting finding was that F-RNA phage groups 1 and 4 were detected along with group 3 in the chicken wastewater. Previously, groups 2 and 3 was found to be more specific to humans and by exception in pigs that were considered to have a similar gastrointestinal physiology and flora to humans partly because of the close living conditions (Hsu *et al.*, 1995; Schaper *et al.*, 2002; Cole *et al.*, 2003). It was therefore, unusual that F-RNA phage group 3 was detected in chicken wastewater on two separate occasions. It is likely that farming activities on the chicken farms could have contributed to human input in the chicken wastewater that was possibly not exclusively chicken waste. The results indicated the specificity of groups 2 and 3 for humans, as these were the only two F-RNA phage groups detected in all four of the hospital wastewater samples.

River sites in the Pietermaritzburg catchment were selected on historical *E. coli* data indicating faecal pollution. Samples from the Baynspruit, Dorpspruit, Slangspruit and uMsunduze rivers were taken between the summer periods of October 2002 and February 2003. The F-RNA analyses were conducted on samples when *E. coli* counts exceeded 10 000.100 mL^{-1} resulting from various faecal inputs. These faecal inputs ranged from the faeces of grazing animals to broken and overflowing blocked sewers and informal sanitation often reaching water sources following heavy rainfall events.

All four F-RNA phage genotypes were detected in the river samples analysed. The plaques were detected as dark circular spots on the X-ray Lumi film and varied in size from small to large. The background colour development was possibly due to a combination of background bacterial growth, inadequate washing, poor handling of the Lumi film and over-exposure. There were instances where poor development of the DNA:RNA hybrids occurred on the film and these membranes were subsequently exposed to the Lumi film for longer periods and this over-exposure was sometimes unavoidable.

In addition, plaques were obtained from selected river water samples, harvested and spotted onto specific areas of Tryptone Yeast Extract Glucose Agar plates to determine the F-RNA phage genotypes present. Up to five environmental samples could be analysed per Tryptone Yeast Extract Glucose Agar plate, saving time and making the genotyping assay for multiple environmental samples more cost effective.

The Slangspruit River situated close to an informal settlement in Pietermaritburg often receives faecal input from run-off and poor sanitation infrastructure unable to contain human sewage. The human faecal input is evident by the detection of F-RNA phage groups 2 and 3 in all samples. Groups 1 and 4 were also detected indicating the presence of animal faecal excrement. Groups 1 and 3 were the only F-RNA phage groups present in samples from 3 different sites on the Baynespruit River sampled on the 23/1/03, whilst all 4 groups were detected from samples taken from the Baynespruit River on other days. This river runs through an industrial area with some cattle grazing and domesticated animals including chickens present in informal settlements. This river often receives faecal input from these sources and the results suggested that both human and animal sources were responsible for the faecal pollution. A similar picture presented in the Dorpspruit River that meanders along the city of Pietermaritburg, even passing horse stables before confluence with the uMsunduze

River. The uMsunduze River that flows through the upper Pietermaritzburg catchment, eventually joins the Umgeni River, receives faecal input from run-off of various animals ranging from cattle and chickens to human faecal matter originating from broken and blocked sewers and informal settlements. Samples taken from various points along the uMsunduze River indicated that there were many instances (n=6) that only groups 2 and 3 were detected suggesting that the main source of faecal pollution in these samples were of human origin. Despite being time consuming and expensive, this technique could be used to help distinguish between human and animal faecal pollution in environmental samples.

The quantification of the sterols detected was done using the gas chromatography results (from HP Chemstation). These results were obtained for the calibration standards that were plotted using Microsoft Excel and calculated from the calibration curve obtained for each batch of samples.

Faecal Sterol Assays on Wastewater Samples

After careful preparation of the 40 ppm, 80 ppm and 160 ppm coprostanol and cholesterol standards, it was injected into the GC. The peaks obtained after approximately 22 min were identified by the mass spectrometer as coprostanol, followed shortly afterwards by the cholesterol peak. The mass spectra of these compounds detected after 22.71 min and 24.33 min were matched to the reference library in the mass spectrometer and confirmed as coprostanol and cholesterol respectively.

The wastewater samples were very turbid and each extraction did not separate as clear chloroform/water phases. All three extracts had to be combined into a separate flask before a proper phase separation could be seen. This additional step was successful in obtaining sterol extracts with minimal trapped water remaining. Following extraction and derivatisation of extracts from the different wastewater samples,. samples were injected into the GC as described above and the mass spectra of peaks obtained after 22 min where identified as coprostanol and cholesterol. These peak areas were quantified once the calibration standards were plotted and the area of abundance under each peak could be converted to the corresponding concentrations. The results indicated that there were instances when higher faecal sterol concentrations were detected (as high as 4 208 $\mu g.L^{-1}$ cholesterol in chicken wastewater) as opposed to instances where relatively low concentrations of 56.3 $\mu g.L^{-1}$ cholesterol was detected in chicken wastewater. These trends of varying low or high concentrations of either cholesterol or coprostanol were found in all of the wastewater samples analysed. It could be attributed to a number of factors including; samples taken over different flows, different diets, microflora in the gut and general health of the individuals.

Coprostanol was generally detected in higher concentrations (23.1 – 1 441 $\mu g.L^{-1}$) than cholesterol (28.2 – 555 $\mu g.L^{-1}$) in both the local hospital and pig wastewater samples. This is possibly due to the gastrointestinal physiology of pigs that have been reported to be similar to that of humans (Scott et al., 2002). The first sample analysed from the local hospital was the only exception where 23.1 $\mu g.L^{-1}$ of coprostanol was detected in the presence of 28.2 $\mu g.L^{-1}$ of cholesterol. Cholesterol was consistently detected at much higher concentrations (181 – 4 208 $\mu g.L^{-1}$) than coprostanol (22.6 – 142.3 $\mu g.L^{-1}$) in both cattle and chicken wastewater. This is in keeping with literature that highlighted that most animals are unable to biohydrogenate cholesterol to coprostanol due to the physiology of the animal gut including the microflora present (Elhmmali et al., 2000).

It is clear that the average concentrations of coprostanol:cholesterol were greater than 2:1 in both human and pig wastewater. The reverse was found with cattle and chicken wastewater where the average cholesterol concentrations were higher (158 – 1 542 $\mu g.L^{-1}$) than the average coprostanol concentrations (49 – 58 $\mu g.L^{-1}$).

Faecal Sterol and F-RNA Phage Genotyping Assays on Environmental Samples

Faecal sterols were generally below the detection limit when 500 mL sample volumes. After numerous attempts it was found that better cholesterol and coprostanol recoveries in environmental samples were obtained using 2 L sample volumes. The results indicated that this method was not sensitive in detecting faecal sterol concentrations below 318 x $10^{-3}\mu g.L^{-1}$. While in general, all 4 genotypes of F-RNA phages were detected in the Msunduze River (site 2), which is downstream of a wastewater treatment plant. The coprostanol concentrations were higher (ranging from 2 424 – 7 107 $\mu g.L^{-1}$) than the cholesterol concentrations (1 592 – 5 670 $\mu g.L^{-1}$). However, the opposite was found in the Dorpspruit River sampled during an agricultural show where the coprostanol concentrations were either below the detection limit or very low while the cholesterol concentrations were as high as (8 014 $\mu g.L^{-1}$). The impact of animal faecal material in the Dorpspruit River was clearly detected by the respective coprostanol:cholesterol ratios and further confirmed by the detection of only group 1 F-RNA phages, known to be of animal origin. The Slangspruit River known to have faecal input from informal settlements in the catchment, contained groups 2 and 3 F-RNA phages known to be mainly of human origin. Apart from confirming the presence of faecal matter in the Slangspruit River, the coprostanol and cholesterol concentrations could not be used to determine if the pollution was of human origin as the coprostanol (17 – 412 $\mu g.L^{-1}$) concentrations detected were lower than the cholesterol concentrations (55 – 1 243 $\mu g.L^{-1)}$. The Baynespruit River contained all 4 groups of F-RNA phages suggesting that the faecal pollution was of human and animal origin but there was 19.1 $\mu g.L^{-1}$ of coprostanol opposed to 9.3 $\mu g.L^{-1}$ of cholesterol suggesting that the faecal matter was predominantly of human origin. This site is known to have faecal pollution resulting from over flowing broken and blocked sewers in the area and would confirm the presumptions made from the faecal sterol assays that the faecal input was predominantly of human origin. Despite the shortcomings in both the F-RNA phage genotyping and faecal sterol assays, both methods could be used to determine the origin of faecal pollution. This could assist with the assessment of associated health risks in the event of faecal contamination of water sources.

Origin of Male Specific F-RNA Bacteriophage Groups in the Primary Water Sources

In the second study genotyping of F-RNA isolates from the communal tap and river water sources for both the villages identified group 1 male specific F-RNA bacteriophages as the predominant bacteriophage group present (Table 2). Group 1 male specific F-RNA bacteriophages are indicative of animal faecal pollutions, specifically cattle, sheep and pig faeces (Hsu *et al.*, 1995; Beekwilder *et al.*, 1996; Uys, 1999). In village 1 using communal tap water sources it was observed that faeces of animals such as pigs, goats and cattle were lying next to the taps. The water reservoir in village 1 was also exposed to small animals, bird droppings and dust particles which might have contained faeces from animals grazing in the vicinity of the reservoir.

The Sambandou River used by households in village 2 was frequently used by domestic animals and cattle for drinking purposes and it was common to find animal faeces in the vicinity of the drinking water sources or close to the areas where people collect their drinking water or even in the water source (Table 2). All these animal activities in the vicinity of the water sources contributed to the presence of group 1 male specific F-RNA bacteriophage contamination that was identified in the water sources. The National Research Council (NRC, 2004) has reported that group 1 male specific F-RNA bacteriophages are found in both human and animals faeces and sewage. Therefore, it could be possible that the predominance of group 1 male specific F-RNA bacteriophages in both the water sources and especially in high concentrations in the river source could be due to both animal and human activities in and near the river source.

No male specific F-RNA bacteriophages belonging to groups 2 and 3 (associated mainly with human faecal pollution) and group 4 (associated mainly with animal faecal pollution) have been isolated from the communal tap water source samples. These bacteriophage groups may have had a fast die-off curve or were just not present at all. A study carried out by Schaper and co-workers (2002b) have shown that group 1 male specific F-RNA isolates were more resistant than group 2 F-RNA isolates followed by group 3 male specific F-RNA isolates and lastly group 4 male specific F-RNA isolates to chlorine, temperature, pH and salt concentrations in water samples. The absence of groups 2, 3 and 4 in the tap water sources could therefore give a false indication that the group 1 isolates were primarily of animal faecal origin and not from human origin (Hsu *et al.*, 1995; Beekwilder *et al.*, 1996; Uys, 1999).

Table 2. Prevalence of male specific F-RNA bacteriophages in river and communal tap water sources in two rural villages in the Vhembe region of the Limpopo Province, South Africa

Village 1 using communal tap water					Village 2 using Sambandou River water				
Male specific F-RNA bacteriophages genotype isolated (percentage %)					Male specific F-RNA bacteriophages genotype isolated (percentage %)				
Number of water samples tested	Group 1 (MS2)	Group 2 (GA)	Group 3 (QB)	Group 4 (F1)	Number of water samples tested	Group 1 (MS2)	Group 2 (GA)	Group 3 (QB)	Group 4 (F1)
14*	14 (100%)	0 (0%)	0 (0%)	0 (0%)	8*	8 (100%)	4 (50%)	0 (0%)	0 (0%)

* Water samples collected for round 1 and round 2

However, male specific F-RNA bacteriophages belonging to group 2 were found in the river water samples (50%), which could indicate possible human pollution of the source water (Hsu *et al.*, 1995; Beekwilder *et al.*, 1996; Uys, 1999; Brion *et al.*, 2002). Group 4 bacteriophages have been shown to be associated with bird faeces (Brion *et al.*, 2002; Schaper *et al.*, 2002a) and even though no group 4 male specific F-RNA bacteriophages were identified during this study, both the river and communal tap reservoirs were exposed to faecal contamination from small animals and birds (Table 2).

Origin of Male Specific F-RNA Bacteriophage Subgroups in the Stored Household Water at the Point-of-Use in the Traditional and CDC Safe Water Storage Containers in Rural Households

A total of four (7%) male specific F-RNA bacteriophages belonging to group 1 male specific F-RNA bacteriophages (associated with animal faecal pollution) was identified in the traditional storage containers in the study households using the tap water source (Table 3). Similarly only one (5%) of the CDC safe storage containers in the households using tap water sources tested positive for the presence of group 1 male specific F-RNA bacteriophages (associated with animal faecal pollution) (Table 3).

In the study households using the river water source, a total of 37 (62%) of the traditional storage containers contained group 1 male specific F-RNA bacteriophages (associated with animal faecal pollution) (Table 3). Similarly, nine (45%) of the CDC safe storage containers tested positive for the presence of group 1 male specific F-RNA bacteriophages (associated with animal faecal pollution) (Table 3). Since animals were observed during this study to lick the communal taps in village 1 and defaecate in the vicinity of the taps and river water area where people collect their domestic water from, the presence of group 1 male specific F-RNA bacteriophages (associated with animal faecal pollution) was similar to the results obtained for the two water sources analysed (Table 2). This is in agreement with similar studies, which found that the presence of group 1 male specific F-RNA bacteriophages in water samples primarily indicated animal faecal pollution (Hsu *et al.*, 1995; Beekwilder *et al.*, 1996; Uys, 1999).

Since it was observed that the storage containers were left out in the yard or stored inside a traditional hut in many instances without a cover, the exposure to dust and faecal contamination from domestic animals, insects and poultry could have introduced group 1 male specific F-RNA bacteriophages to the containers (Rosas *et al.*, 2006). Many of these households also used fresh cow dung to smear the floors of their huts. The dust that originates from the dried cow dung could have contributed to the contamination of the open water storage containers (Benenson, 1995; Rosas *et al.*, 2006).

No group 2 male specific F-RNA bacteriophages (associated with human faecal pollution) were isolated from the tap water sources (Table 2), or in any of the storage containers in village 1 households (Table 3). However, group 2 male specific F-RNA bacteriophages (associated with human faecal pollution) were isolated in both the traditional and the CDC safe storage containers in households from village 2 (Table 3). Nine (15%) of the sixty households (40 households from round 1 water collection and 20 households from round 2 water collections) using the traditional water storage containers and nine (45%) of the twenty households (from second water collection trip) using the CDC safe storage containers contained group 2 male specific F-RNA bacteriophages associated with human pollution (Table 3).

Brion and co-workers (2002) have stated that the presence of group 2 male specific F-RNA bacteriophages was an indication of distant or sporadic faecal pollution of human origin. Previous studies have confirmed that group 2 male specific F-RNA bacteriophages are predominantly found in human faeces and sewage (Hsu *et al.*, 1995; Beekwilder *et al.*, 1996; Uys, 1999). Consequently, contamination of the stored water in this study by humans might have occurred when members of the households used dirty utensils to transfer the stored water from these large open storage containers to a smaller storage container or directly through faecally contaminated hands – especially by small children touching the storage containers and utensils (Jagals *et al.*, 1999). In addition a study conducted in South Africa and Spain (Schaper *et al.*, 2002a) analysed various sewage and faecal samples and showed that faeces from poultry, cattle and pigs could also contribute to the presence of group 2 male specific F-RNA bacteriophages.

Table 3. Prevalence of male specific F-RNA bacteriophages in stored drinking water containers from rural households in two villages in the Vhembe region of the Limpopo Province, South Africa

Household storage container	Village 1 using communal tap water Male specific F-RNA genotypes isolated (percentage %)				Village 2 using Sambandou River water Male specific F-RNA genotypes isolated (percentage %)			
	Group 1 (MS2)	Group 2 (GA)	Group 3 (QB)	Group 4 (F1)	Group 1 (MS2)	Group 2 (GA)	Group 3 (QB)	Group 4 (F1)
Traditional storage containers (n = 60)*	4 (7%)	0 (0%)	0 (0%)	0 (0%)	37 (62%)	9 (15%)	0 (0%)	0 (0%)
CDC safe storage containers (n = 20)**	1 (5%)	0 (0%)	0 (0%)	0 (0%)	9 (45%)	9 (45%)	0 (0%)	0 (0%)

*40 households selected in each village (first water collection round) using traditional storage containers + 20 households (second water collection round) used as control group in each village using traditional storage containers

**20 households selected in each village (second water collection round) using CDC safe storage containers

A close human to animal association were observed in these rural communities and domestic animals and poultry were frequently seen walking into the household area where the water containers were stored. Consequently, the presence of the group 2 male specific F-RNA bacteriophages in the traditional and especially in the CDC safe storage containers suggested that faecal contamination could also have originated from these domestic animals and cattle at the households as well as from the primary water sources (Jagals *et al.*, 1999; Schaper *et al.*, 2002a). No group 2 (associated mainly with human faecal pollution) or group 4 (associated mainly with animal faecal pollution such as poultry and pig faeces) (Schaper *et al.*, 2002a) were detected in any of the traditional or CDC safe storage containers during the study period (Table 3). These results were similar to the results obtained for the primary water sources (Table 3). However, according to a 2004 review on *Indicators for Waterborne Pathogens* by the National Research Council of the National Academies of Science (NRC, 2004), group 1 was found in both human and animals faeces and sewage. Therefore, the absence of group 3 and 4 from the water samples tested during this study, might mean that isolates belonging to these two groups might not persist in water as long as groups 1 and 2 (Schaper *et al.*, 2002b). Consequently, groups 1 and 2 isolates present in these water samples might have been introduced into the water due to both human and animal contamination (NRC, 2004).

Summary of the Use of Male Specific F-RNA Bacteriophages Group Typing to Determine the Faecal Source Origin in Primary Water Sources and Drinking Water Stored in Traditional and CDC Safe Storage Containers in Rural Households

This is the first study to use male specific F-RNA bacteriophages to determine the origin of faecal pollution in household storage containers in rural households without adequate water and sanitation infrastructures. The results demonstrated that water from the water sources and the household storage containers were primarily contaminated by animal faecal matter because mainly group 1 F-RNA bacteriophages (associated with animal faecal pollution) were isolated. In addition, households using an unprotected water source also had group 2 male specific F-RNA bacteriophages present in the household stored water which could have been either due to poor sanitation and hygienic conditions during storage and handling or due to contamination by animal faeces (Rosas *et al.*, 2006). It was difficult to determine the reason for the human faecal contamination because this study did not focus on household hygiene practices. It was however, observed that people removed the taps and the caps of the containers because they were afraid the children would break them. This happened in spite of the educational intervention on the proper use of the CDC safe storage container. In addition a recent study suggests that group 2 male specific F-RNA bacteriophages could have been from faecal samples of poultry and cattle (Schaper *et al.*, 2002a). Consequently, it could be speculated that both groups 1 and 2 isolates could have been introduced to the stored drinking water from both human and animal origin (NRC, 2004). However, it is important to note that Schaper and co-workers (2002a) concluded that the association between the specific groups cannot be used for absolute distinction between human and animal faecal pollution. Genotyping, therefore, seems not to be such an accurate tool to determine the origin of faecal pollution due to the potential for cross-reactions between some human and animal groups (NRC, 2004). This indicated the need for more intensive studies to confirm the specificity of the four groups of male specific F-RNA bacteriophages.

The absence of groups 3 and 4 male specific F-RNA bacteriophage isolates in both the sources and storage containers indicated (i) no human contamination of the household stored water, (ii) isolates from these subgroups does not survive for long periods in the environment and (iii) temperature, pH and turbidity of the water could affect the survival of this specific group isolates (Schaper *et al.*, 2002b). More studies are therefore needed to investigate the prevalence of male specific F-RNA groups in human and animal faeces especially in rural communities where a close living relationship exists between humans and animals.

Although the CDC safe storage container was specifically designed to reduce external microbial pollution of stored drinking water, it was observed that the households did not, at all times put the caps and/or the taps/spigot on the CDC safe storage containers exposing the water in these containers to potential faecal pollution. Possible reasons were that the parents were scared that the children would break the tap or through away the cap, which could have increased the risk of faecal contamination of the water. Although this study reported on a small study group, the results clearly illustrated the need to provide these households with proper water and sanitation infrastructures to reduce the storage period of household drinking water and in the process try to prevent the possible faecal contamination of the stored water. In general this study has found that the use of male specific F-RNA bacteriophage genotyping could be used to some extend to distinguish between human and animal faecal pollution. However, this is an expensive technique which requires skilled personnel and more studies in rural settings are needed. This was however the first study according to the literature to describe the origin of faecal pollution in household stored drinking water in a rural setting.

CONCLUSION

The first study was conclusive in proving that genotyping of F-RNA phages using chemiluminescent detection techniques was easy to perform producing results that were easy to interpret. The use of a hot air oven held at 80°C was successful in fixing the viral RNA onto the positively charged membranes due to a lack of a UV Transilluminator. The lowest background development on the X-ray film was achieved if plaques were transferred before any re-growth of bacteria occurred on plates stored in the cold room (4°C – 10°C). The probes selected for the hydridisation were sensitive and no cross-reactivity between the four F-RNA phages groups occurred.

The results confirmed earlier reports on the specificity of F-RNA phage genotypes 1 and 4 for animal wastes, and genotypes 2 and 3 for human excreta, in a part of the world where not much research has been done (Havelaar *et al.*, 1986). Male specific F-RNA phage group 3 was detected in both chicken and pig wastewater indicating that this group may not be specific to humans. There is a possibility that the chicken wastewater could have been contaminated with human faecal input due to farming activities. The major sources of faecal contamination in river water samples were not established in most instances as all four groups were detected in river water samples. The rivers that were surveyed were influenced by human and animal activities and genotyping used in isolation was not sufficient in determining which of these sources had more impact. There would appear to be an abundance of F-RNA phage genotypes 1 and 2 in the environmental samples, possibly due to their resistance to the prevailing environmental conditions. This research suggested that the results of F-RNA phage genotyping of F-RNA phages isolated from river water samples needs to be used together with other discriminating microbiological or chemical analyses to be more conclusive in establishing whether faecal input was of human or animal origin. Information on the human or animal origin of faecal pollution is important because it gives an indication of the possible risk of infection and the possible water treatment that may be needed to control the transmission of potential waterborne diseases.

The GC-MS method used for the detection of cholesterol and coprostanol was successful and most positive results were obtained when faecal sterol extractions from samples were done as soon as possible after sampling. The average concentrations of coprostanol detected in human (299 $\mu g.L^{-1}$) and pig (623 $\mu g.L^{-1}$) wastewater was higher than their respective average cholesterol concentrations (148 $\mu g.L^{-1}$ and 276 $\mu g.L^{-1}$). The opposite was found with chicken and cattle wastewater where the average cholesterol concentrations (1 542 $\mu g.L^{-1}$ and 158 $\mu g.L^{-1}$ respectively) were higher than the coprostanol concentrations (58 $\mu g.L^{-1}$ and 49 $\mu g.L^{-1}$respectively). The results for coprostanol and cholesterol are in keeping with other research findings where animals in general, apart from pigs, have higher cholesterol than coprostanol concentrations because of the limitations in the gut where the flora present in most animals are unable to biohydrogenate cholesterol to coprostanol (Elhmmali *et al.*, 2000). The analyses were unable to discriminate between human and pig faecal pollution although it has been reported that pigs would have lower concentrations of coprostanol than humans but it would be difficult to discriminate between human and pig faecal pollution in water. When these analyses were applied to environmental samples, larger samples volumes of up to 2 L were used and sterol concentrations as low as 318 x 10^{-3} $\mu g.L^{-1}$ was detected.

Genotyping of F-RNA phages isolated in these environmental samples could sometimes be used to source track faecal pollution, especially when groups associated with either animal or human were identified. Only group 1 F-RNA phages, known to be of animal origin was detected in samples taken from the Dorpspruit during an agricultural show and the specificity of these phages for animal faecal pollution was confirmed by the higher cholesterol:coprostanol ratios obtained. Genotyping results were not useful when all 4 groups of F-RNA phages were detected but the coprostanol and cholesterol concentrations obtained for these sites were more discriminatory in tracking the faecal pollution to be of human or animal origin depending on the respective coprostanol and cholesterol concentrations.

Both these chemical and microbiological methods could be used to determine if pollution was of human or animal origin but more conclusive and reliable discriminatory results were obtained if used together. This was because faecal sterol assays could be used in situations when F-RNA phage genotypes could not be used to distinguish faecal pollution. Only two faecal sterols, predominant in faeces, were selected for the investigation in this study due to the cost implications and specialised chemical methods required for assays of all the sterol assays present in faeces. Analysing for all of the faecal sterols and stanols may provide more useful and conclusive data when the sensitivity of using copostanol and cholesterol alone fails. These chemical and microbial assays are difficult, expensive and time-consuming. Despite having these shortcomings, important and useful data were obtained by applying both chemical and F-RNA phage genotyping assays.

The second study emphasizes the importance to determine the pathways of faecal contamination within the domestic household to decide on an effective point-of-use treatment system. Male specific F-RNA bacteriophage groups were used in this study to determine the origin of faecal pollution. The study was carried out specifically in rural households with a close living association with domestic animals and cattle. Differences in male specific F-RNA bacteriophages prevalence in the storage containers of households using different water sources were seen in this study. The prevalence of male specific F-RNA bacteriophages ranged between 30% and 65% for households using tap water and between 85% and 90% for households using river water. The higher prevalence of phages in the river water could have been due to animal and human activities in or near the river source. In addition, no difference between the traditional and CDC safe storage container water samples were seen with regard to the prevalence of male specific F-RNA bacteriophages. This is in agreement with the results from the formal intervention study (results not included here) indicating that the container without a sodium hypochlorite treatment is not improving the microbiological quality of the stored drinking water.

The results further demonstrated that water from the communal tap water and the household storage containers in village 1 were primarily contaminated by animal faecal matter because the majority of samples contained group 1 male specific F-RNA bacteriophages (associated with animal faecal pollution). However, water from the river water sources and the household storage containers in village 2 were primarily contaminated by animal and human faecal matter because the samples contained group 1 male specific F-RNA bacteriophages (associated with animal faecal pollution) and group 2 male specific F-RNA bacteriophages (associated with human faecal pollution). The results furthermore confirmed that storage containers do not prevent faecal contamination of stored drinking water in the absence of improved hygiene and sanitation behavior practices by the household members.

Since this was the first study to use male specific F-RNA bacteriophages to determine the origin of faecal pollution in household storage containers in rural households; the following aspects were identified for further research:

(1) Survival of male specific F-RNA subgroups in households water storage containers. It would be important to investigate factors such as container type, storage conditions, role of temperature, pH and turbidity, water type, prevalence and role of biofilm in container and the survival period of different male specific F-RNA subgroups
(2) Compare male specific F-RNA subgroups genotyping with new molecular PCR technique (Ogorzaly and Zantzer, 2006) to compare effectivity and costs
(3) Determine the male specific F-RNA subgroups present in different animals from these rural communities to assess the specificity of male specific F-RNA subgroups typing as a source tracking technique.

Consequently the results of the two studies did give some indication of the origin of faecal pollution. Research has shown that there is no single method that could be used with absolute certainty and therefore, a combination of detection and identification methods could contribute to a better understanding in an attempt to obtain more useful and meaningful data regarding the origin of faecal pollution. Additional studies may be needed to confirm and develop these methodologies.

Although people know that the water can be contaminated, they are often ignorant of their contribution to faecal pollution at the resource as well as point-of use level. Educational aspects need to be addressed focusing on health issues, which will include training on basic hygiene and sanitation practices at the household level. Research on all the various levels will be of extreme importance in water quality studies and will provide valuable data to improve the microbial quality of water both at the source and at the point-of use.. New data will assist various role players and decision makers at government level in the formulation of policies regarding water, sanitation and hygiene aspects and changes in countries such as South Africa to improve the general well-being of its people.

AKNOWLEDGEMENTS

The authors would like to thank the Department of Medical Virology, University of Pretoria/NHLS and Umgeni Water for analyses conducted.

REFERENCES

Allwood PB, Malik YS, Hedberg CW, Goyal SM. Survival of F-specific coliphage, feline calicivirus, and *Escherichia coli* in water: a comparative study. *Applied and Environmental Microbiology* 2003;69:5707-5710.

Armon R. Bacteriophage monitoring in drinking water: do they fulfil the index or indicator function? *Water Science and Technology* 1993;27:463-470.

Beekwilder J, Nieuwenhuizen R, Havelaar AH, Van Duin J. An oligonucleotide hybridization assay for the identification and enumeration of F-specific RNA bacteriophages in surface water. *Journal of Applied Bacteriology* 1996;80:179-86.

Benenson. AS. Control of communicable diseases in man. 16th ed. Washington: American Public Health Association; 1995. p 530.

Bligh EG, Dyer WJ. A rapid method of total lipid extraction and purification. *Canadian Journal of Biochemistry and Physiology* 1959;37:911–917.

Brion GM, Meschke JS, Sobsey MD. F-specific RNA coliphages: Occurrence, types, and survival in natural waters. *Water Research* 2002;36:2419-2425.

Brown RC, Wade TL. Sedimentary coprostanol and hydrocarbon distribution adjacent to a sewage outfall. *Water Research* 1984;18:621–632.

Calci KR, Burkhardt 3rd W, Watkins WD, Rippey SR. Occurrence of male - specific bacteriophage in feral and domestic animal wastes, human faeces, and human - associated wastewaters. *Applied and Environmental Microbiology* 1998;64:,5027-5029.

Chan KH, Lam MHW, Poon KF, Yeung HY, Chiu TKT. Application of sedimentary fecal stanols and sterols in tracing sewage pollution in coastal waters. *Water Research* 1998;32:225–235.

Christie WW. Lipid Analysis: Isolation, Separation, Identification and Structural Analysis of Lipids. 2nd ed. Oxford: Pergamon Press; 1982.

Cole D, Long SC, Sobsey MD. Evaluation of F+RNA and DNA coliphages as source-specific indicators of fecal contamination in surface waters. *Applied and Environmental Microbiology* 2003;69:6507–6514.

Debartolomeis J, Cabelli VJ. Evaluation of an *Escherichia coli* host strain for enumeration of F male - specific bacteriophages. *Applied and Environmental Microbiology* 1991;57:1301–1305.

Donnison AM, Ross CM. Somatic and F - specific coliphages in New Zealand waste treatment lagoons. *Water Research* 1995;29:1105–1110.

Department of Water Affairs and Forestry (DWAF). *Holmes ed. South African Water Quality Guidelines : Domestic Use.* 1996;1:89–90.

Dutka BJ, Chau ASY, Coburn J. Relationship between bacterial indicators of water pollution and faecal sterols. *Water Research* 1974;8:1047–1055.

Elhmmali MM, Roberts DJ, Evershed RP. Combined analysis of bile acids and sterols/stanols from riverine particulates to assess sewage discharges and other fecal sources. *Environmental Science and Technology* 2000;34:39–46.

Fowles IA. Gas Chromatography. 2nd ed. Chichester: John Wiley and Sons; 1998.

Furniss BS, Hannaford AJ, Smith PWG, Tatchell AR. Vogels. Textbook of Practical Organic Chemistry. 5th ed. New York: John Wiley and Sons, Inc; 1989.

Furuse K. Distribution of coliphages in the environment: general considerations. In: Goyal SM, Gerba CP, Bitton, G, editors. Phage Ecology. New York: Wiley – *Interscience*; 1987. p. 87–124.

Geldenhys JC, Pretorius PD. The occurrence of enteric viruses in polluted water, correlation to indicator organisms and factors influencing their numbers. *Water Science and Technology* 1989;21:105–109.

Gerba CP, Rose JB, Haas CN, Crabtree KD. Waterborne rotavirus: A risk assessment. *Water, Science and Technology* 1996;30:2929-2940.

Grabow WOK, Coubrough P, Nupen EM, Bateman BW. Evaluation of coliphages as indicators of the virological quality of sewage - polluted water. *Water SA* 1984;10:7–14.

Grabow WOK. Waterborne diseases: Update on water quality assessment and control. *Water SA* 1996;22:193-202.

Grabow WOK, Very A, Uys M, de Villiers JC. Evaluation of the application of bcateriophages as indicators of water quality. South African Water Research Commission Final Report 540/1/98. 1997.

Graham Solomons TW. Organic Chemistry. 2nd ed. USA: John Wiley and Sons, Inc: 1980.

Griffin DW, Stokes R, Rose JB, Paul JH. Bacterial indicator occurrence and the use of an F^+ specific RNA coliphage assay to identify fecal sources in Homosassa Springs, Florida. *Microbial Ecology* 2000;39:56–64.

Gundry S, Wright J, Conroy R. A systematic review of the health outcomes related to household water quality in developing countries. *Journal of Water and Health* 2004;2:1-13.

Havelaar AH, Hogeboom WM. A method for enumeration of male -specific bacteriophages in sewage. *Journal of Applied Bacteriology* 1984;56:439–447.

Havelaar AH, Furuse K, Hogeboom WM. Bacteriophages and indicator bacteria in human and animal faeces. *Journal of Applied Bacteriology* 1986;60:255–262.

Havelaar AH, Pot-Hogeboom WM, Furuse K, Pot R, Hormann MP. F - Specific RNA bacteriophages and sensitive host strains in faeces and wastewater of human and animal origin. *Journal of Applied Bacteriology* 1990;69:30–37.

Havelaar AH, Van Olphen M, Drost YC. F-specific RNA bacteriophages are adequate model organisms for enteric viruses in fresh water. *Applied and Environmental Microbiology* 1993;59:2956-2962.

Hsu FC, Shieh YSC, Van Duin J, Beekwilder MJ, Sobsey MD. Genotyping male - specific RNA coliphages by hybridisation with oligonucleotide probes. *Applied and Environmental Microbiology* 1995;61:3960–3966.

Hurst CJ, Benton WH, Stetler RE. Detecting viruses in water. *Journal of the American Water Works Association: Research and Technology* 1989;9:71–80.

International Association for Water Pollution Research and Control (IAWPRC). Study group on Health related Water Microbiology. AH Havelaar ed. Bacteriophages as model viruses in water quality control. *Water Research* 1991;25:529–545.

ISO 10705–1. Water quality: Detection and enumeration of bacteriophages. Part 1: Enumeration of F-specific RNA bacteriophages, International Organisation for Standardisation. 1995.

Isobe KO, Tarao M, Chiem NH, Minh LY, Takada H. Effect of environmental factors on the relationship between concentrations of coprostanol and fecal indicator bacteria in tropical (Mekong Delta) and temperate (Tokyo) freshwaters. *Applied and Environmental Microbiology* 2004;70:814–821.

Jagals P, Grabow WOK, De Villiers JC. Evaluation of indicators for assessment of human and animal faecal pollution of surface run-off. Water, *Science and Technology* 1995;31:235-241.

Jagals P, Bokato TC, Grabow WOK. Changing consumer water-use patterns and their effect on microbiological water quality as a result of an engineering intervention. *Water SA* 1999;25:297-300.

Kirkwood A. Safe water for Africa. DFID Conference Features. Africa Health 1998:9-11.

Kosek M, Bern C, Guerrant RL. The global burden of diarrhoeal disease, as estimated from studies published between 1992 and 2000. Bulletin of the World Health Organization 2003;81:197-204.

Leclerc H, Schwartzbrod L, Dei-Cas E. Microbial agents associated with waterborne diseases. *Critical Reviews in Microbiology* 2002;28:371-409.

Leeming R, Ball A, Ashbolt N, Nichols P. Using faecal sterols from humans and animals to distinguish faecal pollution in receiving waters. *Water Research* 1996;30:2893–2900.

Leeming R, Nichols PD. Concentrations of coprostanol to existing bacterial indicator guideline limits. *Water Research* 1996;30:2997–3006.

Leeming R, Bate N, Hewlett R, Nichols PD. Discriminating faecal pollution : a case study of stormwater entering Port Phillip Bay, Australia. *Water Science and Technology* 1998;38:15–22.

Lin J, Biyda PT, Puckree T, Bezuidenhout CFC. Antibiotic resistance profiles and environmental isolates from Mhlathuze River, KwaZulu-Natal (RSA). Water SA 2004;30:23-28.

Macdonald IA, Bokkenheuser JW, McLernon AM, Mosbach EH. Degradation of steroids in the human gut. *Journal of Lipid Research* 1983;24:675–694.

Martin WJ, Ravi Subbiah MT, Kottke BA, Birk CC, Taylor M C. Nature of faecal sterols and intestinal bacterial flora. *Lipids* 1973;8:208–215.

Matusik JE, Hoskin GP, Sphon JA. Gas Chromatography/Mass Spectrometric confirmation of identity of coprostanol in Mercenaria mercenaria (Bivalvia) taken from sewage-polluted water. *Journal of the Association of Official Analytical Chemists* 1988;71:994–999.

McCalley DV, Cooke M, Nickless G. Effect of sewage treatment on faecal sterols. *Water Research* 1981;15:1019–1025.

Moyo S, Wright J, Ndamba J, Gundr S. Realising the maximum health benefit from water quality improvements in the home: A case study from Zaka District, Zimbabwe. *Physics and Chemistry of the Earth* 2004;29:1295-1299.

Murtaugh JJ, Bunch RL. Sterols as a measure of fecal pollution. *Journal Water Pollution Control Federation* 1967;39:404–409.

Nguyen DK, Bruchet A, Arpino P. Determination of sterols in sewage sludge by combined *in situ* trimethylsilylation/supercritical fluid extraction and GC-MS. *Environmental Science and Technology* 1995;29:1686–1690.

Nichols PD, Leeming R, Rayner MS, Latham V. Use of capillary gas chromatography for measuring fecal - derived sterols. Application to stormwater, the sea - surface microlayer, beach greases, regional studies, and distinguishing algal blooms and human and non - human sources of sewage pollution. *Journal of Chromatography* 1996;733:497–509.

NRC. Indicators for waterborne pathogens. Committee on Indicators for Waterborne Pathogens, National Research Council. Internet URL: http://www.nap. edu/catalog/ 11010. html. 2004.

Obi CL, Potgieter N, Bessong PO, Matsuang G. Scope of potential bacterial agents of diarrhoea and microbial assessment of quality of river water sources in rural Venda communities in South Africa. Water, *Science and Technology* 2003;47:59-64.

Obi CL, Green E, Bessong PO, De Villiers B, Hoosen AA, Igumbor EO, Potgieter N. Gene encoding virulence markers among *Escherichia coli* isolates from diarrhoic stool samples and river sources in rural Venda communities of South Africa. *Water SA* 2004;30:37-42.

Ogorzaly L, Zanther C. The development of real time RT-PCR for specific detection of F-specific RNA bacteriophage genotypes. *Journal of Virological Methods* 2006;138:131-139.

Osawa S, Furuse K, Watanabe I. Distribution of ribonucleic acid coliphages in animals. *Applied and Environmental Microbiology* 1981;41: 64–168.

Pool FP. Recent advances in the Silylation of organic compounds for gas chromatography. In: Blau K, King GS editors. Handbook of Derivatives for Chromatography. London: Heydon;1978. p. 152–200.

Rhodes MW, Kator HI. Use of *Salmonella typhimurium* WG 49 to enumerate male - specific coliphages in an estuary and watershed subject to non - point pollution. *Water Research* 1991;25:1315–1323.

Rosas I, Salinas E, Martinez L, Calva E, Cravioto A, Eslava C, Amabile-Cuevas CF. Urban dust fecal pollution in Mexico City: Antibiotic resitance and virulence factors of *Escherichia coli.* International Journal of Hygiene and Environmental Health 2006;209:461-470.

Schaper M, Jofre J. Comparison of methods for detecting genotypes of F - specific RNA bacteriophages and fingerprinting the origin of faecal pollution in water samples. *Journal of Virological Methods* 2000;89:1–10.

Schaper M, Jofre J, Uys M, Grabow WOK. Distribution of genotypes of F-specific RNA bacteriophages in human and non-human sources of faecal pollution in South Africa and Spain. *Journal of Applied Microbiology* 2002a;92:657-667.

Schaper M, Duran AE, Jofre J. Comparative resistance of phage isolates of four genotypes of F-specific RNA bacteriophages to various inactivation processes. *Applied and Environmental Microbiology* 2002b;68:3702-3707.

Scott TM, Rose JB, Jenkins TM, Farrah SR, Lukasik J. Microbial source tracking: current methodology and future directions. *Applied and Environmental Microbiology* 2002;68:5796–5803.

Sinton LW, Finley RK, Hannah DJ. Distinguishing human from animal faecal contamination in water: A review. New Zealand *Journal of Marine and Fresh Water Research* 1998;32:323-348.

Skraber S, Gassilloud B, Schwartzbrod L, Gantzer C. Survival of infectious poliovirus-1 in river water compared to the persistence of somatic coliphages, thermotolerant coliforms and poliovirus-1 genome. *Water Research* 2004;38:2927-2933.

Sobsey MD, Dufour AP, Gerba CP, LeChavallier MW, Payment P. Using a conceptual framework for assessing risks to health from microbes in drinking water. *Journal of the American Water Works Association* 1993;85:44-48.

Sobsey MD. Managing water in the home: Accelerated health gains from improved water supply. World Health Organization Sustainable Development and Healthy Environments. World Health Organization, Geneva. WHO/SDE/WSH/02.07. 2002.

Theron J, Cloete TE. Emerging waterborne infections: contributing factors, agents and detection tools. *Critical Reviews in Microbiology* 2002;28:1-26.

UN. United Nations Millennium Declaration. General Assembly Res. 55/2. 18 September 2000.

Uys M. MSc Dissertation: *Molecular characterisation of F-specific RNA phages in South Africa.* Department of Medical Virology, University of Pretoria, South Africa. 1999.

Venkatesan MI, Santiago CA. Sterols in ocean sediments : novel tracers to examine habitats of cetaceans, pinnipeds, penguins and humans. *Marine Biology* 1989;102:431–437.

Vivian CMG. Tracers of sewage sludge in the marine environment: a review. *The Science of the Total Environment* 1986;53:5–40.

Watson JT. Introduction to mass spectrometry. 2nd ed. New York: Raven Press; 1985.

WHO. The world health report: Reducing risks, promoting healthy life. World Health Organization, Geneva. Internet URL: http://www.who.int/whr/en. 2002a.

WHO. Water for Development: A practical advocacy guide for World Water Day 2002. Internet URL: http://www.worldwaterday.org/advocacy/adv.html. 2002b.

WHO. Heterotrophic plate count measurement in drinking water safety management, Protection of the Human environment. Water, Sanitation and Health. World Health Organization, Geneva. WHO/SDE/WSH/02.10. 2002c.

Writer JH, Leenheer JA, Barber LB, Amy GL, Chapra SC. Sewage contamination in the upper Mississippi river as measured by the fecal sterol, coprostanol. *Water Research* 1995;29:1427–1436.

In: Aquatic Toxicology Research Focus
Editor: Elias P. Svensson, pp. 79-98

ISBN: 978-1-60456-192-0

Chapter 4

SIDE-EFFECT OF PESTICIDES ON AQUATIC ORGANISMS

Shao-nan Li*[3] *and Sheng-li Zhou
Institute of Pesticide and Environmental Toxicology, College of Agriculture and Biotechnology, Zhejiang University, Hangzhou 310029, China

ABSTRACT

This paper reviewed impacts of currently used pesticides to aquatic organisms including fish, amphibian, aquatic invertebrates and algae in terms of acute, longer-term, joint effects, and side-effect on ecological levels. Pesticides showed diversity in their toxicity to aquatic species. Among them organophosphates and pyrethroids were more toxic to aquatic animals than the other pesticides, They were estimated to be moderately to highly toxic in most cases. Due to ubiquitous inter-specific diversity in sensitive, dubious relationship between acute and longer-term toxicity, and various direction of joint effect, a correct and complete risk assessment was hard to achieve by limited data. Sublethal concentration of pesticides might alter physical or biochemical parameters, impair self-protecting system, or change behavior of exposed aquatic species. Pesticides also caused alteration in species configuration and energy transfer in aquatic systems. Researches cited in this paper indicated that some of the pesticides could pose high risk to fish or invertebrates, even when they were applied at authorized rate or recommended dosage.

INTRODUCTION

Different from other anthropogenetic pollutants pesticides are chemicals introduced deliberately by human being to the environment. Annual consumption of pesticides in China was estimated to be 22~26 tons in last 10 years (Marked Trends, 2006). Among the whole amount of pesticides, 54.6% were insecticides, 19.3% were herbicides, and 25.3% were fungicides (Expert Review, 2002). Application rate of pesticide is quite different among

[3] Corresponding author: Tel. 0086 571 86971946; E-mail snli@zju.edu.cn

regions in China. According to Cai (1999), the rate was 2.33kg/ha in average, with the lowest rate of 0.36kg/ha in Inner Mongolia, and the highest rate of 9.96kg/ha in Zhejiang province.

Pesticides, as they are introduced to the Environment, have chances to contaminate surface water in ways of direct spraying, spray drift, run-off, and so on, and thus make hazards to various kinds of aquatic organisms which reside in paddy field or nearby surface water.

Aquatic ecosystem includes a large number of species they can roughly be divided into aquatic vertebrate (fish and amphibian), aquatic macroinvertebrate (*e.g.,* shrimp, crab, aquatic insect, and so on), aquatic plant (emerged, floating, and submerged), zooplankton (*e.g.,* protozoan, rotifer, cladocera, and copepod, and so on), phytoplankton (cyanobacteria, bacillariophyta and chlorophyta) and so on. Commonly, algae are chose as representative species of producer, water flea (cladoceras) as primary consumer and fish as secondary consumer in aquatic ecosystem.

Although fish is on the top of food web in aquatic ecosystem and holds higher economic value, it doesn't mean that other species weigh less and need less attention to be protected. In fact, being affected by pesticides at any linkages of the food chain always leads to unwanted consequence to aquatic ecosystem.

Side effect of pesticides on aquatic organisms was previously reviewed by some researchers, such as, Zhang (1988), who reviewed influence of pyrethroid pesticides on aquatic ecosystems, and Yang et al. (2002), who reviewed progress on research of toxic effect of organophosphorous pesticides on aquatic organism. The present paper reviewed toxicity and sub-lethal effect of currently used pesticides on major groups of aquatic organisms. Since a pesticide nowadays seldom appears as a single in aquatic environment. Joint effect and change in toxicity as a pesticide combined with other organic and inorganic pollutants was also reviewed. In addition to that attention was also paid on ecological-leveled responses and reaction of various aquatic organisms as they were exposed to the recommended dosage of pesticides.

1. TOXICITY OF PESTICIDES TO AQUATIC ORGANISMS

Pesticides can impose both acute and longer-term hazards to aquatic organisms residing in agricultural ecosystem according to use pattern and physicochemical characters of pesticides in concern, and to biological characters and behavior of aquatic organisms involved. Generally speaking, a pesticide which is repeatedly applied and is stable in water is more possible to impose longer-term hazard to aquatic organisms than a pesticide which is applied only once a season and is unstable in water. From organism point of view, those who reside in sediment have more chance to receive longer-term poisoning from pesticides exposure.

1.1. Acute Toxicity

Acute toxicity, as it is mentioned, refers to stresses after a short term exposure of pesticides in high dosage. The same as in general toxicology, acute toxicity of pesticides to aquatic organisms is usually scaled with LD_{50}, LC_{50}, or EC_{50}, and so on.

1.1.1. Fish

In assessing hazardous potential to fish, pesticides were classified into 3 categories according to the value of 96h LC_{50} (Cai et al., 1987):

- LC_{50} < 0. 1ppm: highly toxic
- LC_{50}: 0. 1~1. 0ppm: moderately toxic
- LC_{50} > 1. 0ppm: slightly toxic

The above division is based on following assumptions:

- The maximal application rate of most pesticides never exceeds 750g per hectare in paddy field.
- In a "fish/rice culture system", some ditches are dug for fish residing.
- Pesticide concentration in the field with "fish ditches" is about 1 ppm immediately after the application.
- A pesticide that causes <50% mortality to the field-residing fish is considered to be slightly toxic.
- A pesticide that causes >50% mortality, due to the run-off contamination, to fish which resides in off-target area, is considered to be highly toxic.

Based on above-mentioned Cai's Division and a safety factor of 10, the Chinese State Environmental Protection Agency put forward an assessment criteria as following (Chinese EPA, 1989):

- LC_{50} <1.0ppm: highly toxic
- LC_{50}: 1.0~10ppm: moderately toxic
- LC_{50} >10ppm: slightly toxic

At present stage, the state regulative department of pesticides of China tends to add one more grade to the assessment system, *i.e.*, "extremely toxic". A pesticide with LC_{50} (96h) value <0.1mg/l is considered "extremely toxic".

In China, goldfish (*Carassius auratus*) and zebra fish (*Danio rerio*) are recommended to serve as representative species for registration-oriented toxicity test. A series of 96h toxicity tests conducted with zebrafish indicated that among the 121 of about 400 pesticide formulations, which were put to registration in 2006 in China, 27.3% and 16.5% of them were highly toxic and extremely toxic, respectively (Table 1 and Figure 1).

Table 1. Percentage distribution in terms of 96h toxicity of pesticides applying for registration in 2006 (n=121)

Pesticides	Percentage distribution				
	Extremely toxic	Highly toxic	Moderately toxic	Slightly toxic	Total
insecticides	15.7	21.5	18.2	6.61	62
herbicides	0	1.65	7.44	8.26	17.4
fungicides	0.826	4.13	6.61	9.09	20.7
Total	16.5	27.3	32.2	24	100

It can be seen from Table 1 that insecticides, as a whole, are much more toxic to fish than the other two groups, *i.e.*, herbicides and fungicides.

Triazophos is an organophosphate insecticide which is widely used for leaf born worm control in paddy field in China. The LC_{50} (96h) of triazophos 20%EC is 0.035, 0.060, and 0.008mg/L, respectively, for freshwater fish *Tilapia nilotica*, *Colossoma brachypomum* and *Peseudorasobora parva* (Li, 1998), and The LC_{50} (48h) of triazophos 20%EC for marine fish *Latealabrox japonicus*, *Anguilla japonica, Boleophthalmus pectinirostris,* and *Liza haematocheila* is 0.082, 0.090, 0.009 and 0.004mg/L, respectively (Jin, 2002). According to the presently accepted criterion, triazophos 20%EC is extremely toxic to both freshwater fish and marine fish mentioned above.

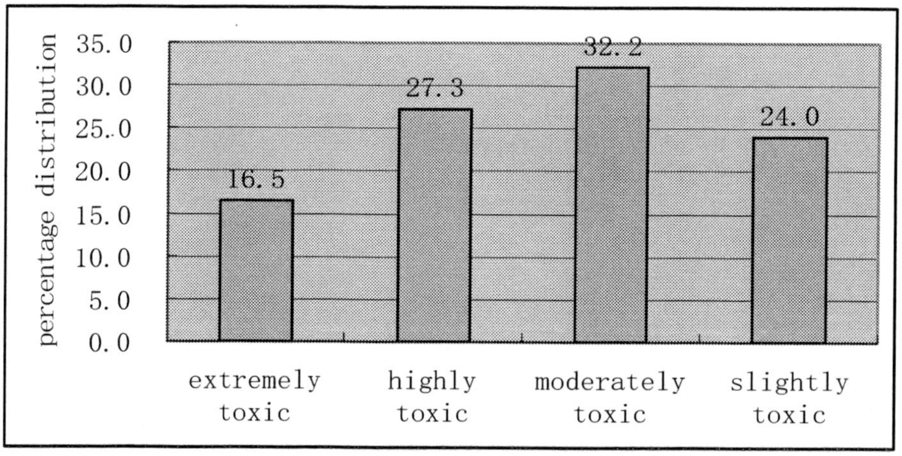

Figure 1. Percentage distribution in terms of acute toxicity of pesticides applying for registration in 2006 (n=121).

Most of the pyrethroid insecticides are found to be extremely toxic to most of the fish species. Wang *et al.* (2000) reviewed acute toxicity of 9 broadly used prethroids to five fish species. Six of the insecticides, *i.e.*, fenpropathrin, deltamethrin, cypermethrin, fenvalerate, *beta*-cypermethrin, and *lambda*-cyhalothrin, which were all with cyano ligand in molecule, had their LC_{50} (48h~96h) fell in range of 0.00012~0.00721mg/L for crucian carp, carp, mosquito fish, and gudgeon (*Cobiocypris rarus*), and LC_{50} (48h) of 0.0106~0.105mg/L for eel, *Paramisgurnus dabryanus*. The other three pyrethroids, *i.e.*, chlovaporthrin, methothrin and tetramethrin, which were all without cyano ligand, were relative low toxic to fish. They had their LC_{50} (48h) fell in range of 0.0325~0.883mg/L for the above-mentioned five species of fish.

Although the zebrafish is recommended by Chinese administrative department and some international organizations, such as, ISO, to serve as representative species for registration-oriented toxicity test, susceptive diversity among different species was ubiquitous. Li and Fan (1996) measured acute toxicity diversity of malathion to topmouth gudgeon (*Pseudorasbora parva*), goldfish (*Carassius auratus*), tilapia (*Oreochromis niloticus*), mosquitofish (*Gambusia affinis*) and rainbow trout (*Oncorhynchus mykiss*). The study indicated pure malathion (>99% in purity) held LC_{50} (96h) of 14.5, 11.3, 4.6, 0.7 and 0.25mg/L, respectively, for the above-mentioned five fish species, with maximal susceptive discrepancy of 60 folds.

Triphenylphosphate (TPP) and piperonyl butoxide (PBO) are known to be specific inhibitor of carboxylesterase and monooxygenase, respectively. In practice, the two compounds usually act as synergist of organophosphorus insecticides and pyrethroid insecticides, respectively. For investigating reasons for species-related difference in susceptibility, Li and Fan (1996) measured toxicity of malathion as the fish were pretreated with TPP and PBO, respectively. Result indicated that TPP enhanced toxicity of malathion by 5~7 times to all the three species of tested fish, *i.e.*, topmouth gudgeon, goldfish, and mosquitofish, while fish species responded differently to PBO. PBO slightly enhanced toxicity of malathion to the species belonging to Cyprinidae, *i.e.*, topmouth gudgeon and goldfish, while it dramatically decreased toxicity of malathion to mosquitofish and rainbow trout, which belongs to Cyprinodontidae and Salmonidae, respectively, and were relatively sensitive to malathion. This suggested that selective toxicity of malathion among fish species resulted from different capacity of them to transfer malathion into its oxon analog.

Another sample for species-related selective toxicity is given by triazophos. Previous researches indicated that this organophosphorus insecticide was extremely toxic to many kinds of species of freshwater and marine fish, with 48h to 96h LC50 in range of 0.004~0.09mg/L as showed by Li (1998) and Jin (2002). The authors found, however, that triazophos of 20%EC, 3%G and 40%EW hled LC_{50} (96h) of 5.22, 5.67, and 6.54mg/L, respectively, to zebrafish (unpublished data). A recent study conducted by Xie *et al.* (2006) indicated that 24th, 48th, 72nd, and 96^{th}h LC50 of triazophos to grass carp, *Ctenopharyngodon idellus,* were 1.93, 1.73, 1.60, and 1.50 mg/L, respectively.

Conclusion can be derived from above examples that considering the inter-specific diversities in sensitivity, correct and complete risk assessment can hardly achieve if the toxicity data is obtained merely from one fish species, even it was recommended as representative of the whole group.

1.1.2. Amphibian

As amphibian resides in aquatic environment, as tadpoles, in the most sensitive state of their lifecycle, this group of animal was also included as representative species by the State Environmental Protection Administration of China. Acute toxicity of pesticides to amphibian was usually evaluated by the same criteria as that set for fish (Chinese EPA, 1989).

Pan and Liang (1996) determined 48 hour toxicity of 20 insecticides, 26 herbicides and 32 fungicides, towards marsh frog tadpole (*Rana limnocharis*). The LC_{50} value of the test was summarized in Table 2.

Table 2. Percentage distribution in terms of 48h toxicity of pesticides to marsh frog tadpole (*Rana limnocharis*) (Pan and Liang , 1996)

Pesticides	Percentage distribution[a]				
	Extremely toxic	Highly toxic	Moderately toxic	Slightly toxic	Total
Insecticides	6.41 (25)[b]	3.85 (15)	6.41 (25)	8.97 (35)	25.6 (100)
Herbicides	0.00 (0)	2.56 (7.69)	7.69 (23.1)	23.1 (69.2)	33.3 (100)
Fungicides	0.00 (0)	3.85 (9.38)	16.7 (40.6)	20.5 (50.0)	41.0 (100)
Total	6.41	10.3	30.8	52.6	100

[a]Highly toxic: LC_{50}<1.0ppm; moderately toxic: 1.0ppm<LC_{50}<10ppm; slightly toxic: LC_{50} > 10ppm.
[b]Inclosed in the blank is percentage distribution against total numbers of each group of pesticides

As showed in table 2, insecticides were more toxic than herbicides and fungicides to tadpole of marsh frog. 25% of the tested insecticides were extremely toxic to marsh frog, with 48 hour LC_{50} values of <0.1ppm, while the proportional number for herbicides and fungicides was zero. 15% of insecticides were highly toxic to the frog, with 48 hour LC_{50} values falling in the range of 0.1 to 1ppm, while the proportional number for fungicides and herbicides was 9.38% and 7.69%, respectively. Among various groups of insecticides the tadpole was relatively sensitive to pyrethroids and organochlorines. All the pyrethroid and organochlorine insecticides tested displayed 48 hour LC_{50} of <1 ppm.

1.1.3. Aquatic Invertebrates

Acute toxicity tests of two herbicides, *i.e.*, oxadiazon and quizalofop-*p*-ethyl, and two insecticides *i.e.*, pymetrozine, and acetamiprid were conducted by Hu *et al.* (2002) in lab using *Daphnia magna* and zebrafish as test animal species. Result showed that 48h EC_{50} to *Daphnia magna* was 1.44mg/L, 1.21mg/L, 14.6mg/L, and 0.06mg/L, respectively, for oxadiazon, quizalofop-*p*-ethyl, pymetrozine, and acetamiprid. As for zebrafish, 96h EC_{50} was 3.09mg/L, 0.91mg/L, 120mg/L, and 0.53mg/L, respectively, for oxadiazon, quizalofop-*p*-ethyl, pymetrozine, and acetamiprid. It seemed that the aquatic crustacean was comparatively sensitive than fish to pesticide exposure.

Zhou *et al.* (2003) tested acute toxicity of 20% triazophos EC to a species of Crustacea *Artemia cysts,* a species of prawn *Penaeus vannamei,* and a species of water snail *Anadara granosa.* The insecticide showed LC_{50} (48h) of 0.800mg/L, 0.0032mg/L and 21.0mg/L, respectively, to the three species of aquatic invertebrates, and the LC_{50} (96h) of 0.0011mg/L and 10.2mg/L was detected, respectively, for the latter two species of aquatic invertebrates. According to the criteria set by the State Environmental Protection Agency of China (Chinese EPA, 1989), 20%Triazophos EC was highly toxic to *A. cysts,*and *P. vannamei*, and it was low toxic to *A. granosa.* Whereas according to the criteria for evaluating of acute toxicity of pesticides to aquatic invertebrates, which was currently adopted by the state regulative department of pesticides of China, 20%Triazophos EC was highly toxic to *A. cysts*, extremely toxic to *P. vannamei*, and low toxic to *A. granosa.*

Du (2001) studied acute toxicity of organophosphorus pesticides to *Artemia sinica.* Result indicated that 48h LC_{50} of phoxim, dichlorvos, methmidophos, omethoate, and parathion was 0.12, 0.20, 54, 180 and 0.52, respectively.

Chen and Chen (1993) tested toxicity of two organophosphorus insecticides, *i.e.*, 40% isofenphos-methyl EC and 40% isocarbophos EC to two species of seawater-snail, *i.e., Mytilus edulis* and *Potamocorbula laevis, and* one species of prawn, *Penaeus chinensis*. The result indicated that LC_{50} (96h) of the two insecticides fell in range of 2.40~7.13mg/L for the two species of seawater-snail, and LC_{50} (20d) of isofenphos-methyl was 0.0000628mg/L for the prawn.

In another study Chen *et al.* (1997) tested toxicity of two pyrethroid insecticides, *i.e.*, 40% fenvalerate EC and 20% tetramethrin EC to two species of seawater-snail, *i.e., Mytilus edulis* and *Chlamys farreri*. The result indicated that LC_{50} (96h) of the two insecticides fell in the range of 0.30～2.34mg/L.

According to Wang *et al.* (2000), the LC_{50} (48h) of 6 cyano- contained pyrethroid, *i.e.*, fenpropathrin, deltamethrin, cypermethrin, fenvalerate, *beta*-cypermethrin and *lambda*-cyhalothrin, to *Daphnia carinata* fell in the range of 0.000069~0.00056mg/L, while the LC_{50} (48h) of chlovaporthrin was calculated to be 0.0398mg/L to the same species.

Shan *et al.* (2003) determined toxicity of 5% fipronil SC, a phenylpyrazole insecticide, to two species of prawn, *i.e. Macrobrachium rosenbergii, Macrobrachium nipponensis*, and one species of crab, *i.e., Eriocheir sinensis*. The result indicated that LC_{50} (96h) of the insecticide was 0.00098, 0.00432 and 0.00856mg/L, respectively, for *Macrobrachium rosenbergii, Macrobrachium nipponensis,* and *Eriocheir sinensis*.

It can be seen from above data that most species from the crustacean were sensitive to insecticide exposure. According to the currently adopted criteria the insecticides involved were highly to extremely toxic tò this group of organisms, except for those from anostraca, *e.g., Artemia sp.,* to which some of the organic phosphorus insecticides were low toxic. The available data also indicated that, among the various groups of aquatic invertebrates, mollusca were relatively insensitive to insecticide exposure. Insecticides involved were moderately toxic to this group of organisms according to the currently adopted criteria.

1.1.4. Alga

Chen and Chen (1993) tested acute toxicity of two orgnophosphorus insecticides, *i.e.*, 40% isofenphos-methyl EC and 40% isocarbophos EC to a species of seawater algae, *Isochrysis galbana*. The result indicated that EC_{50} (96h) was 0.66mg/L and 2.03mg/L, respectively, for isofenphos-methyl and isocarbophos.

In another study Chen *et al.* (1997) tested acute toxicity of two pyrethroid insecticides, *i.e.*, 40% fenvalerate EC and 20% tetramethrin EC to two species of seawater algae, *i.e., Isochrysis galbana* and *Chlorella spp.* The EC_{50} (96h) of the two insecticides fell in the range of 0.52~1.38mg/L.

According to Tang and Li (1999), the EC_{50} (72h) of three orgnophosphorus insecticides, *i.e.*, monocrotophos, parathion, and phoxim to a seawater alga, *Phaeodactylum tricornutum*, was calculated to be 9.74mg/L, 8.20mg/L and 1.52mg/L, respectively.

Based on a study performed by Wang *et al.*, the (1999) 48h and 96h EC_{50} of six orgnophosphorus insecticides, *i.e.*, phoxim, parathion, omethoate, methamidphos, dechlorvos, and isocardophos to a species of freshwater algae, *Chlorella vulgaris*, was calculated to be 1.25 and 1.12mg/L, 7.0 and 6.0mg/L, 90.0 and 57.0mg/L, 15.8 and 11.7mg/L, 13.4 and 6.5mg/L, and 92.3 and 104.5mg/L respectively.

Acute toxicity tests of two insecticides, *i.e.*, pymetrozine and acetamiprid, were conducted by Hu *et al.* (2002) in lab using freshwater algae, *Scenedesmus obliquus*, as test organism. Result showed that 96h EC_{50} was 1.61mg/L and 4.18mg/L, respectively, for pymetrozine, and acetamiprid.

Zhou *et al.* (2003) tested the acute toxicity of 20% triazophos EC to *Chlorella pyrenoidosa*. The result indicated that LC_{50} (48h) of the insecticide was 0.024mg/L.

Zhang and Cai (1986) proposed criteria, based on 5d (120 hour) EC_{50} values, for evaluating toxicity of pesticides to algae. The criteria are as following:

- $EC_{50}<0.3$mg/L high toxicity
- $0.3<EC_{50}<3.0$mg/L moderate toxicity
- $EC_{50}>3.0$mg/L low toxicity

This classification is based on following assumptions:

- Depth of water in paddy field is usually kept for 5 cm.
- Application rate of a pesticide never exceeds1500g per hectare.
- Pesticide concentration in the field, in this case, is about 3.0 ppm.
- The final concentration of pesticides in adjacent water, due to run-off, is about 1/10 of that in the field.
- A pesticide that causes <50% growth inhibition to field residing algae is considered slightly toxic.
- A pesticide that causes >50% inhibition, due to the run-off contamination, to the growth of algae in off-field water column, is considered highly toxic.

The state regulative department of pesticides of China tends to accept the criteria, whereas duration of the test is shortened from 5d to 96h.

Based on above-mentioned data and currently adopted criteria one can see that most of the insecticides were slightly toxic to moderately toxic, except for triazophos, which was highly toxic, to algae, while toxicity of herbicides or fungicides was versatile due to the possible differences of mode of action of different groups.

Acute toxicity tests of two herbicides, *i.e.*, oxadiazon and quizalofop-*p*-ethyl, were conducted by Hu *et al.* (2002) in lab using freshwater algae, *Scenedesmus obliquus,* as test organism. Result showed that 96h EC_{50} was 0.01mg/L and 0.01mg/L, respectively, for oxadiazon and quizalofop-*p*-ethyl.

Acute toxicity of three kinds of herbicides, i.e., cinmethylinon, oxyfluorfen, and fluazifop-butyl, on *Scenedesmus obliquus* was measured by Deng (2000). Result showed that 96h EC_{50} were 18.4mg/L, 0.0005mg/L, and 5.37mg/L, respectively, for cinmethylin, oxyfluorfen, and fluazifop-butyl.

Relative sensitivity of green algae, *Chlorella vulgaris*, to three commonly used herbicides, metribuzin, fenoxaprop-*p*-ethyl, and alachlor, was determined by Yang *et al.* (2004). Result indicated that 96h EC_{50} was 0.021mg/L, 0.937mg/L, and 5.54 mg/L, respectively, for metribuzin, fenoxaprop-*p*-ethyl, and alachlor.

Acute toxicity (96h EC_{50}) of a novel sulfonylurea herbicide, HNPC-C9908, to green algae, *Chlorella pyrenoidosa*, was calculated to be 29.1mg/L (Ou *et al.*, 2003).

Acute toxicity (96h EC_{50}) of paclobutrazol, a plant growth regulator, was 41.5mg/L, to *Selenastrum capricornutum* (Cai *et al.*, 1994).

Based on above-mentioned criteria one can see that some of the herbicides were highly toxic, the others were moderately and even slightly toxic to algae. This diversity might result from differences in terms of mode of action of different groups of herbicides.

1.2. Longer-Term Toxicity

Different from acute toxicity, which relates to short-term exposed at high dosage, longer-term toxicity refers to responses under continuous exposure, such as 10~30% proportion of the lifecycle, or even the whole or several generations. It is not rare to see a pesticide which is acutely nontoxic may lead to longer-term consequences in case of continuous exposure.

According to Zheng *et al.* (1999), LC_{50} (48h) of 20% fenpropathrin EC was 0.011mg/L to crucian carp (*Carassius auratus).* If the fish was exposed for 60 days, the highest no effective concentration (NOEC) and the lowest effective concentration (LOEC), as it was counted with body weight, was 0.0007mg/L and 0.0014mg/L, respectively. The maximal acceptable concentration (MATC) was therefore estimated to be 0.0010mg/L, which was lower by one order of magnitude than the corresponding LC_{50} (48h).

Paclobutrazol, a plant growth regulator with 48h LC_{50} of 33.2mg/L to *Daphnia magna*, was considered a pesticide of low toxic. A 21d experiment conducted by Yang *et al.* (2003) indicated that 21d LC_{50} of paclobutrazol was 1.14mg/L, and NOEC of the pesticide was 0.75mg/L. Further experiment indicated that if the offspring was exposed to pesticide concentration of 0.75mg/L mortality of 63% and 83% would be detected, respectively, for generation of F_1 and F_2. This suggested that the offspring was more sensitive than the parent to pesticide exposure.

In a study conducted by Tan *et al.* (2004), chronic toxicity of fenvalerate, *beta*-cypermethrin, chlorpyrifos, and triazophos towards water flea, *Daphnia magna*, was tested in a period of 21d. Result indicated that parameters, such as, "number of offspring for first brood", "total number of offspring", "total number of broods", and "intrinsic rate of population increase" seemed more sensitive than parameters, such as, "days to first pregnancy" and "days to first breeding" to insecticide exposure. It was found that body length of daphnia was obviously affected in fenvalerate and cypermethrin exposure, while the body length was not affected in chlorpyrifos and triazophos exposure, so the body length might be able to serve as a specific index for stress of pyrithroid insecticides. Selecting the most sensitive paramenter as the index, NOEC was calculated to be 0.005, 0,005, 0.01, and 0.05ug/L, and LOEC was 0.010, 0,010, 0.05, and 0.10ug/L, respectively, for fenvalerate, *beta*-cypermethrin, chlorpyrifos, and triazophos.

Considering the time and expense needed for longer-term toxicity tests people long have the intention to predict longer-term responses from acute toxicity tests. For this purpose, a concept of so called "application factor (AF)" was introduced to reflect relations between acute LC_{50} or EC_{50} and longer-term NOEC or LOEC. As showed in study performed by Tan *et al.* (2004) 48h LC_{50} was measure to be 0.61, 0.55, 7.77, and 13.8ug/L, respectively, for fenvalerate, *beta*-cypermethrin, chlorpyrifos, and triazophos. AF value of the 4 insecticides was calculated to be 0.0123, 0.0136, 0.00386, and 0.00543, respectively, indicating that the AF values were quite different for the insecticides involved.

1.3. Joint Toxicity

It is reasonable to image that pesticides rarely occur in aquatic environment as a single compound, because more and more pesticides and other anthropogenetic compounds were introduced. Inherent toxicity of pesticides may change as they contact each other or with other environmental pollutants. The pesticide-related joint toxicity can be additive, synergistic, and antagonistic. Since combination between pesticides and between pesticide and other pollutants is ubiquitous in aquatic environment more attention should be paid on joint effects resulted from the combinations.

Xie *et al.* (1999) studied joint toxicity of some organophosphate and heavy metal to three species of ocean algae, *i.e.*, *Phaeodactylum tricornutum*, *Dunaliella sp.* and *Platymonas sp.* In amount proportion of 1:1 toxic value of 40% Profenofos EC and copper, the additive index (AI) was found to be -0.462, -0.557 and -0.702 respectively, suggesting that the mixture of profenofos and copper was antagonistic to all the three species of algae.

Dai *et al.* (1997) studied joint toxicity of oranophosphorus insecticides and heavy metals to seawater fish, *Pagrosomus major* and *Rhabdosargus sarba.* Result indicated that copper was antagonistic to toxicity of isofenphos-methyl to *Pagrosomus major* and *Rhabdosargus sarba*, while toxicity of isofenphos-methyl potentially increased in existence of methamidophos.

In study of joint toxicity of petroleum pollutant, anthracene, and organophosphorus insecticide, Wang *et al.* (2000) found that anthracene significantly increased toxicity of profenofos to three species of ocean algae, *i.e.*, *Phaeodactylum tricornutum*, *Dunaliella sp.* and *Platymonas sp.*

Ru *et al.* (1996) measured joint toxicity of 4 organophosphorus to seawater alga, *Platymonas sp.*. Result showed that monocrotophos and phoxim were synergistic in their toxicity to *Platymonas sp.*, while monocrotophos and dichlorvos were antagonistic. Mixtures of monocrotophos+parathion and dichlorvos+parathion were found to be synergistic at 48th h, but they were antagonistic at 72nd h and 96th h, respectively.

Xie *et al.*, (2006) measured joint toxicity of organophosphorus insecticides and pyrethroid insecticide to fish. They found that cypermethrin and triazophos were antagonistic in their toxicity to grass carp, *Ctenopharyngodon idellus,* at 24th, 48th, 72nd, and 96^{th}h of the exposure. As for cypermethrin and phoxim, the two insecticides were found to be synergistic at 24^{th}h, and antagonistic at 48th, 72nd, and 96^{th}h of the exposure, respectively.

Du (2001) studied joint toxicity of organophosphorus pesticides to *Artemia sinica.* Result indicated that 48h toxicity of phoxim and dichlorvos, methmidophos and omethoate, and parathion and phoxim was antagonistic, with AI value of -1.67, -1.50, and 6.38, respectively.

Above data indicated that interaction between pesticides themselves and between pesticides and other organic or inorganic pollutants, in terms of toxicity to aquatic organisms, was ubiquitous. As showed in this review, direction of the interaction (synergism or antagonism) was related, not only to chemical belongings of pesticides and environmental pollutants involved, but also the time span of exposure. Also, the direction and degree of the interaction seemed difficult to predict.

Qian *et al.* (2000) estimated joint toxicity of anti- cholinesterase insecticides to carp based on inhibition of brain acetylcholinesterase (AchE). The *in vitro* test indicated that the combing parathion with omethoate, methamidophos, and aldicarb, respectively, all gained additive effects, and the combination between the oganophosphate and carbamate insceticide had more obviously synergistic effect than the combinations between the organophosphates.

Gu *et al.* (1999, 2000) estimated joint effect of an organophosphorus insecticide, malathion, and a phenylpyrazole insecticide, fipronil, whose primary target was reported to be on chloride channel of neuron cells, on brain AchE of topmouth gudgeon. Result indicated that fipronil had no inhibition on brain AchE. Whereas after the fish was translated to freshwater, recovery of brain AchE was much delayed for fish samples that were co-exposed to fipronil and malathion compared with those that were merely exposed to malathion.

2. Action of Pesticides on Aquatic Organisms

2.1. Enzymatic Responses

Mosquitofish (*Gambusia affinis*) were exposed to sublethal concentrations (0.4 mg/L to 1.3mg/L) of an organophosphorus insecticide, fenitrothion, in a static system, and activity of brain acetylcholinesterase (AchE) and hepatic glutathione S-transferase (GSTase) was evaluated after 24, 48 and 96h exposure (Sorsa *et al.* 1999; 2000a). The result indicated that fenitrothion produced significant inhibitory effect, ranging from 44.2% to 65.7%, to brain AchE and 9.4% to 33.5% to GSTase. Recovery of brain AchE was also assessed in fish which previously had been exposed for 96h in fenitrothion (Sorsa *et al.* 1999). Result indicated that following 8 days of recovery period, the AchE activity resumed to 65.8% to 85.85 of the original level. This suggested that AchE depression caused by fenitrothion exposure required long period to recover.

In addition to the mosquitofish Sorsa *et al.* investigated *in vivo* inhibition of brain AchE (2000b) and GSTase (2000a) to another species of freshwater fish, topmouth gudgeon (*Pseudorasobora parva*), after the fish were exposed to sublethal concentrations of fenitrothion (0.2mg/L to 0.6mg/L) in static system. The result indicated that fenitrothion produced AchE inhibition of 9.1% to 54.9% and GSTase inhibition of 1.9%~66.9% as the fish were exposed for a period of 96h. The study indicated that following 8 days of recovery period, the AchE activity of the exposed fish resumed 68.6% to 106.7% of the original level (Sorsa *et al.*, 2000b).

Fish brain AchE has long been considered as a potential biomarker for water contamination caused by anticholinesterase insecticides, *e.g.*, organophosphates and carbamates. Efficiency of the biomarker is restrained, however, by fluctuation of the baseline resulting from variations in inherent state or ambient conditions. In a study performed by Li *et al.* (2005), fluctuation of AchE content of topmouth gudgeon (*Pseudorasbora parva*), as it was exposed to different concentration of triazophos (0.00010mg/L to 0.048mg/L), was measured using an indirect non-competitive ELISA which was established against polycolonic antibodies developed from AchE of the same species. Result indicated that symptoms of poisoning were obvious as the fish was exposed to triazophos of 0.006mg/L, and motality occurred at 0.012mg/L. The maximal rate of induction towards content of brain

AchE reached up to 67.8%. Over the whole period of exposure of 10d, inhibition of brain AchE was in range of -10.9%~91.9%, and the *in vivo* inhibition could only be detected at low concentration of 0.00010mg/L in case activity of AchE was scaled with enzyme content. While in case that activity of AchE was scaled with total protein content the inhibition was detected in range of -25.9%~90.1%, and the minimal detection limit towards the inhibition rose up to 0.0016mg/L.

2.2. Oxidative Damage

In the long history of evolution, organisms have developed anti-oxidative system, including enzymatic and non-enzymatic components, such as GPx, POD, SOD, GSH, CAR, and so on. Under normal physical condition, reactive oxygen generated in metabolism can be eliminated efficiently by this system to keep a balance between generation and elimination. But under pesticide stress, the system may be damaged, and as a consequence the reactive oxygen will accumulate in body and harm the cells.

In a study performed by Tang and Li (1999), reactive oxygen (superoxide anion radicals) level was found to be increased, and lipid peroxidation and esterification to be enhanced, as seawater alga, *Phaeodactylum tricornutum*, was exposed, at a concentration of corresponding EC_{50} (72h), to monocrotophos, parathion, and phoxim, respectively.

Hazard of organophosphorus insecticides to anti-oxidative system of algae was also detected by many other researchers, such as, Xie *et al.* (2000) and Tang *et al.* (1995a; 1995b; 1996; 2001). In spite of that, one should keep in mind that the above-mentioned side-effect could only achieve when the algae were exposed to so high concentrations of several to ten mg/L or more. It is clear that, unless in accident or so, the algae have few possibilities to be exposed, in natural bodies of water, to so high concentrations of organophosphorus insecticides. In contrary, problems of organophosphorus insecticides might reside in their stimulation to alga growth at environmental relevant low concentration (Liu Mei, Liu Zan 1998).

2.3. Avoidance and Adaption

Gong (1994) studied avoidant reaction of fish to 4 pesticides, *i.e.*, propachlor, pyrimyxthion, methyl isofenphos, and monocarboxaldehyde. The study indicated that the carp (Cyprinus carpio) displayed obvious reaction of avoidance to propachlor, pyrimyxthion, and methyl isofenphos. Detected concentration of avoidance was 0.02, 0.086, 0.100mg/L, respectively, which was about 1/10 to 1/20 of their corresponding LC_{50}. Monocarboxaldehyde, however, had obvious paralysis effect to the fish. The fish became stagnant and lost avoidant ability as soon as the carp touched monocarboxaldehyde. Considering avoidant reaction of fish some of the pesticides, such as, propachlor, pyrimyxthion, and methyl isofenphos, might not be so toxic as indicated by the values of LC_{50}, whereas the other one, i.e., monocarboxaldehyde, might be especially dangerous to fish.

In addition to fish, avoidance was also observed in terms of swamp shrimps and lobsters as they contact with pyrethroid insecticide deltamethrin (Cai *et al.*, 1997).

In a study conducted by Ye *et al.* (2004) , water flea *Daphnia magna* was exposed for 21 days to sublethal concentrations of a pyrethroid insecticide bifenthrin. The offspring of the first and third brood, *i.e.*, $F_I(1^{St})$ and $F_I(3^{rd})$, were then transferred to a pesticide-free medium for recovery. During a 21-day recovery period, number of offspring per female and body length was found to reduce significantly for the $F_I(1^{St})$, while all biological parameters for the $F_I(3^{rd})$ restored, except for those whose parents were exposed to bifenthrin of higher than 0.5ug/L . For these $F_I(3^{rd})$ the body length was still obviously affected. This result suggested that the parent daphnia had already adapted, at certain extent, to the pesticide exposure.

3. Impact of Pesticide on Aquatic Organisms at Ecological Level

Impact of pesticide on aquatic ecosystem was determined not only in sensitivity of aquatic organisms, but also in residual level of pesticides in aquatic environment. Zhang (1999) estimated possible impact of pesticides in the sea area of Amoy city in Fujian province based on available literatures on side effects and contamination of pesticides. Conclusions were gained as following: 1) because of reduction in environmental import, hazard caused by organochlorine insecticides was not essential, though they held relatively high toxicity to aquatic organisms; 2) herbicides also makes little harm to sea area due to their moderately high toxicity and their relatively small quantity of consumption; 3) bactericides, fungicides, and carbamate insecticides may cause moderately risk to sea area due to their moderate toxicity and moderate quantity of consumption. 4) organophosphorus insecticides had large quantity of consumption, strong water solubility and high toxicity, thereby they might cause high risk to sea area.

Nie and Huang (1997) investigated side effect of a insecticide monoamitraz to aquatic organisms in system level. Five 60cm×30cm×40cm water/sediment model ecosystems were constructed which contained different species of plant and animal, including macrophyte, phytoplankton, zooplankton, zoobenthos, and a large amount of microorganism. The insecticide was applied at dosages of 1.5mg/L and 3.0mg/L, respectively. Observation indicated that the aquatic vascular plants couldn't stand the exposure. Their leaves turn to yellow as they contacted the insecticide. During the exposure diversity index in terms of the community of plankton was found to be reduced. The exposure resulted in an increase in proportion of *Cyanophyceae*, *Chlorophyta* and *Euglenophyceae* and a reduction in proportion of *Diatomacae,* while *Cryptophyceae*, *Chrysophyceae*, *Pyrrophyceae* and *Xanthophyceae* seemed extinct. Monoamitraz reduced diversity index of zooplankton as well. It was found that monoamitraz obviously changed metabolic function of the model ecosystems from autotrophic to heterotrophic.

Xu *et al.* (2002) reviewed system-level responses of experimental lake ecosystems to three chemical stresses (acidification, copper and pesticide contamination) using exergy and structural exergy as ecological indicators. The pesticides involved included, 2,4 D-DMA, pentachlorophenol (PCP), atrazine, hexachlorobiphenil (HCBP), hexazinone, permethrin, bifenthrin, Carbaryl, chlorpyrifos and theirs metabolite, trichlorophenol (TCP), and so on. According to the article pesticides in lake ecosystem might enlarge cell size of phytoplankton, diminish body size of zooplankton, increase biomass level of phytoplankton and zooplankton, reduce species richness or diversity, and decrease biomass ratios of zooplankton to phytoplankton and zooplankton to zooplankton. Changes were also observed in ecological function in terms of decreases in primary production of phytoplankton, in carbon assimilation ratios for both phytoplankton and zooplankton, in efficiency of ecological transfer, and in ratios for gross production and respiration, and so on. Macro-zooplankton was found to be the most sensitive community, followed by micro-zooplankton and then by phytoplankton. Direction of change of biomass of phytoplankton depended on the toxicity of the pesticides in concern. The research indicated that under pesticide stress, daphnia extincted, while small cladoceran, such as *Bosmina,* and the smallest zooplankton, rotifers, took the dominant place of zooplankton community. It is known that the small cladoceran and rotifers have lower filtering rates and they can only gaze particles of narrow size spectrum. In addition to that the small-size zooplankton is less vulnerable to predation by planktivorous fish. As a result the over-growth of small cladoceran and rotifers in pesticide stress led to frantically reproduction of algal population and finally resulted in deterioration of water quality.

Following are some examples that demonstrated impact of pesticides, as they were applied in authorized rate, to aquatic species in paddy field and in vicinal pond.

For investigating side-effect of pesticide to organisms in paddy field and pound ecosystem, Cai *et al.* (1997) sprayed 2.5% deltamethrin EC to a full-growing rice field at dosage of 7.5g a.i./ha, and two hours later after the spray, two in three of the surface water in the field was ejected to a vicinal pound. The consequences were recorded as follows: 1) After receiving deltamethrin, death rate was detected to be 6.7%, 6.7%, and 0, respectively, for grass carp *Ctenopharyngodon idella*, carp *Cyprinus carpio,* and bighead silver carp *Aristichthys nobilis,* which resided in surface water of target area. No fish and mussel which resided in the vicinal pound were found to be affected. 2) High death rate was found for swamp shrimp and lobster which were kept in the cages in the vicinal pound, with mortality of 86.6%, 80.0%, and 10.0% for the swamp shrimps, and 83.3%, 66.6%, and 23.3% for the lobsters, depending on distance of the organisms from outfall of the paddy field. 3) Zooplankton showed distinguishable response to deltamethrin exposure, with the most severe response in terms of *Cladocera*, followed by *Copepoda* and protozoa, and little response was found in terms of rotifer.

In a middle sized paddy field/pound artificial ecosystem, 3% Isazofos G was applied at dosage of 0.045g a.i./m^2. The surface water in the field was then ejected to a vicinal pound four hours after the pesticide application (Shan et al., 2001). Side effects of the pesticide were recorded as follows: 1) In early time of the exposure, mortality was found in the field in terms of crucian carp *Carassius auratus,* silver carp *Hypophthalmichthys molitrix,* grass carp *Ctenopharyngodon idellus*, carp *Cyprinus carpio* and loach *Paramisgurnus dabryanus.* Mortality of above-mentioned fish was about 45% in average, and the rate reached up to 80% for the loach 48h after the exposure. The survived fish showed obvious symptom as well. 2) Fish in fish-ditches was less severely affected, with death rate of 15% in average 24h after the exposure. 3) Grass shrimp was more sensitive than fish. All the shrimp in the ditches was found died at 24h after the exposure. 4) No fish was found died in the vicinal pound, but high death rate (85%) could still be detected at the end of the trial.

Ultizing the same artificial system, a trial was conducted by Shan *et al.* (2002) with 5% Fipronil SC as test substance. The insecticide was applied to the field at a recommended dosage of 45g a.i./ha^2. One in five of surface water was ejected into the vicinal pound 24h after the spray. Observation indicated that pond-resided prawn and crab displayed symptom of poisoning immediately after the field water was released, and large quantity of the prawn and crab died 1~2d after the field water was released. The time for which total mortality occurred was at 3^{rd}d, 7^{th}d, and 10^{th}d of exposure, respectively for two species of prawn, *i.e. Macrobrachium rosenbergii, Macrobrachium nipponensis*, and one species of crab, *i.e.*, *Eriocheir sinensis*.

In summary, some of the insecticides might pose high risk to fish and aquatic invertebrate, which resided in target area, although the insecticides were applied at recommended dosages. Compared with fish macro-invertebrates, *e.g.*, shrimp and crab, were more sensitive to insecticide exposure. The data indicated that not only the ones in target area, but those in vicinal area, were affected as the field water was ejected to the surrounding ponds after the insecticide application.

4. CONCLUSION

This paper reviews the ecotoxicity of recently used pesticides to aquatic organisms including fish, amphibian, aquatic invertebrates and algae.

One could see from available data that insecticides, as a whole, were more toxic than herbicides and fungicides to fish. Among major groups of insecticides organophosphates displayed variety in their toxicity: A major proportion was moderately or highly toxic, and a small proportion was extremely toxic to fish. In the contrary, pyrethroid insecticides, as a whole, were extremely toxic, to fish.

Although some species of fish, *e.g.*, zebrafish, are selected by Chinese administrative department and by other national or international organizations, such as ISO, to serve as representative of the whole group in susceptibility evaluation, one should keep in mind that susceptive diversity among species was ubiquitous, therefore, correct and complete risk assessment could hardly achieve by data from one or two fish species.

Toxicity of pesticides to amphibian is not so thoroughly studied as that of fish. In spite of that one could see from limited data that insecticides, as a whole, were more toxic than herbicides and fungicides to this group of animal.

Aquatic invertebrates include various groups of organisms. Available data indicated that most of the crustacean, such as those from decapoda (shrimp and crab) and those from cladocera (daphnia) was sensitive to insecticide exposure. The available data also indicated that mollusca was relatively insensitive to insecticide exposure.

Based on currently adopted criteria one can see that most of the insecticides were slightly toxic to moderately toxic to algae, while toxicity of herbicides was quite different due to the possible differences in terms of mode of action of different groups of herbicides.

Pesticides may cause either acute or longer-term hazards to aquatic organisms. According to the available data, the values of NOEC or LOEC measured from longer-term toxicity tests were usually lower by less than 1 to more than 2 orders of magnitude than the corresponding 48h or 96h LC_{50} or EC_{50}. This difference might resulted from diversities either on physicochemical or biochemical characters of pesticides involved, or on physiological or biochemical characters of aquatic organisms involved.

Since more and more pesticides and anthropogenetic compounds were introduced in environment, pesticides have many chances to meet with each other and with other pollutants in aquatic environment. Available data indicated that interaction between pesticides themselves and between pesticides and other organic or inorganic pollutants, in terms of toxicity to aquatic organisms, was ubiquitous. As showed by available data that direction of the interaction (synergism or antagonism) was related not only to chemical belongings of pesticides and environmental pollutants involved, but also to the time span of exposure. Also, the direction and degree of the interaction seemed difficult to predict.

According to available data pesticide residue might resulted in ecological consequence such as fluctuation of species diversity, alteration in dominant species, and change in ecological function, such as decreasing in primary production, in efficiency of ecological transfer, in ratios for gross production and respiration, and so on.

Available data indicated that some of the insecticides might pose high risk to fish and aquatic invertebrates, which resided in target area; even when the insecticides were applied at recommended dosage. Compared with fish, macro-invertebrates, *e.g.*, shrimp and crab, were more vulnerable to insecticide exposure. Not only the ones in target area, but also those in non-target area, were affected as the field water was ejected to the surrounding ponds after the insecticide application.

Although pesticide risk assessments achieved, in limited cases, by field experiment or trials conducted in simulated ecosystems, most of the assessment was based on short-term experiment which was conducted in lab with single species of fish. for registration-oriented toxicity test.

REFERENCES

Marked Trends Analysis. (2006). Market requirement of pesticides in China is possible to increase at rate of 8% annually. *The New Century of Agrochem*, (8): 18.

Expert Review: (2002). Market requirement of pesticides in China are estimated to be 257,000 tons. *China Petroleum and Chemical Industries*, (3): 35~36.

Cai, Dao-ji. (1999). Pesticides and Environment. In Dao-ji Cai (ed.), *Researche methods on Pesticide Environmental Toxicology* (First edition, p.7). Beijing: Publishing Company of Environmental Science of China.

Zhang, Xiao-qun. (1988). Influence of pyrethroid pesticides on aquatic ecosystems. *Journal of Shanghai Jiaotong University (Agricultural Science)*, 6(1): 51~66.

Yang, Xian-le; Zhan, Jia; Huang, Yan-ping. (2002). Progress on research of toxic effect of organophosphorous pesticides on aquatic organism. *Journal of Shanghai Fisheries University*, 11(4): 378~382

Ye, Wei-hong; Liu, Wei-ping; Tan, Ya-jun. (2004). Daphnia magna Toxicity and Subsequent Recovery During a Subchronic Toxicity Test with Bifenthrin. *Chinese Joumal of Pesticides*, 43(2): 86~89.

Chinese Environmental Protection Administration (EPA). (1989) Guideline for Safety Assessment of Chemical Pesticides.

Cai, Dao-ji; Wang, Li-jing. (1987) Researches on ecological safety assessment for chemical pesticdes, IX. A primary research and assessment on toxicity of chemical pesticides to fish. *Rural Eco-environment*, (2): 7~11.

Li, Shao-nan. (1998) Acute toxicity of isofenphos-methyl and triazophos to 3 kinds of fresh water fish. *Environmental Pollution and Control*, 20(3): 15~16.

Jin, Cai-xing; Ding, Yue-ping; Liu, Shi-zhong; Guo, Yuan-ming; Zhu, Jing-ping; Zhong, Zhi. (2002) Acute toxicity of triazophos to perch and other several species of fish. *Fishery Science and Technology Information*, 29(4): 156~158.

Wang, Zhao-Hui; Yin, Yi-wei; Lin, Xiao-tao; Luo, Yu-min; Lin, Qiu-qi; Xu, Zhong-neng. (2000) A review on toxicity of pyrethroid pesticides to aquatic ecosystems. *Journal of Jinan University (Natural Science)*, 21(3): 123~127.

Sorsa, S. S.; Li, Shao-nan; Fan, De-fang. (1999). Brain acetylcholinesterase response and recovery in mosquitofish (*Gambusia affinis*) exposed to fenitrothion. *Chinese Journal of Pesticide Science*, 1(2) : 61~67.

Sorsa, S. S.; Li, Shao-nan; Fan, De-fang. (2000b) In Vivo Inhibition and Recovery of Brain Acetylcholinesterase in Topmouth Gudgeon (*Pseudorasobora parva*) following Exposure to Fenitrothion. *Journal of Zhejiang University (Science)*, 1(4): 448~455.

Sorsa, S. S.; Fan, De-fang; Li, Shao-nan. (2000a) Hepatic Glutathione S-Transferase Activity in Mosquitofish (*Gambusia affinis)* and Topmouth Gudgeon (*Pseudorasobora* parva) Exposed to Fenitrothion. *Journal of Zhejiang University (Science)*, 1(2): 190~195.

Hu, Shuang-qing; Yin, Da-qiang; Chen, Liang-yan. (2002). Safety evaluation of four new pesticides in aquatic ecosystem. *Rural Eco-Environmen*, 18(4): 23~26, 3.

Ou, Xiao-ming; Lei, Man-xiang; Huang, Ming-zhi;, Wang, Yue-long; Wang, Xiao-guang; Fan De-fang. (2003). Effects of novel sulfonylurea herbicide HNPC-C9908 on growth of green algae *Chlorella pyrenoidosa. Chinese Journal of Pesticide Science*, 5(3): 16~23.

Deng, Zhao-hui; Yuan, Jun-feng; Chen, De-hui; Zuo, Ben-tong. (2000). The Effects of Three Kinds of Herbicides on *Scenedesmus obliquus. Journal of Shanghai Teachers University (Natural Sciences),* 29(1): 92~95.

Wang, Cui-hong; Xu, Jian-hong; Xing, Xiao-yun; Fan, Wei; Ma, Xiao-yong; Yang, Li-hong; Zhang, Shu-rong. (1999). Comparative study of the toxicity of six organophosphrous pesticides and three heavy metal ions to *Chlorella vulgaris. Henan Science*, Suppl: 108~110.

Li, Shao-nan; Xie Xian-chuan; Tan Ya-jun; Zhu Guo-nian. (2005). Induction of Triazophos to Brain Acetylcholinesterase from topmouth gudgeon, *Pseudorasbora parva. Chinese Journal of Pesticide Science*, 7(1): 59-62.

Yang, Zhi-qiang; Dong, Bo; Wu, Jin-cai. (2004). Sensitivity of Chlorella vulgaris to metribuzin, puma and alachlor. *Chinese Journal of Applied Ecology*, 15(9): 1621~1625.

Cai, Hou-jian; Zhou, Feng-fan; Jin, Qi; Chen, Zhong-lin; Yu, Ang; Jin, Hong-jun; Huang, Hua-qiang. (1994). Study of PP333 toxicities to several aquatic organisms and effects on plant ultra-microstructures. *Journal of Nanjing University (Natural Science Edition)*, 30(2): 274~280.

Gong, Rui-zhong. (1994). Study on avoidance reaction of fish to four pesticides. *China Environmental Science*, 14(5): 381~384.

Dai, Jia-yin; Deng, Wei-yun; Wang, Shu-hong. (1997). A research on joint toxicity of heavy metal and organophosphorus pesticides to *Pagrosomus major* and *Rhabdosargus sarba. Environmental Science*, 18(5): 44~46, 54.

Wang, You; Tang, Xue-xi; Li Yong-qi. (2000) The joint toxicity of anthracene and profenofos on marine microalga. *Marine Science*, 24(4): 5~7.

Ru, Shao-guo; Li, Yong-qi; Yuan Jun-feng; Jing, Yong-chang. (1996). Studies on the joint toxicity of organophosphorus pesticides to *Platymonas sp.*. *Journal of Ocean University of Qingdao*, 26(2): 197~202.

Xie, Wen-ping; Ma, Guang-zhi; Lai, Zi-ni. (2006). Acute and joint toxicity of cypermethrin and organophosphorus pesticides to juvenile grass carp, *Ctenopharyngodon idellus*. *Water Conservancy and Fishery*, 26(1): 98~103

Du, Juan. (2001) The study of joint toxicity organic phosphorus pesticide to artemia. *Jiangsu Environmental Science*, 14(1): 9~10.

Li, Shao-nan; Fan, De-fang. (1996). Correlation between biochemical parameters and susceptibility of freshwater fish to malathion. Journal of Toxicology and Environmental Health, 48: 413~419.

Chen, Bi-juan; Chen, Min-shan. (1993). The toxic effect of isofenphos-methyl and isocarbophos on marine algae (*isochrysis galbana*), seashell (*Mytilus edulis* and *Potamocorbula laevis*), and juvenile shrimp. *Marine Environmental Science*, 12(2): 12~17.

Zhou, Chang-yi; Chi, Xin-cai, Huang, Cheng; Chen, Wei-quan; Yan, Chong-ling. (2003) Study on toxicity and safety evalution of triazophos on four species of hydrophytic organisms. *Journal of Oceanography in Taiwan Strait*, 22(3): 319~324.

Shan, Z; Wang, L.; Cai, D.; Gong, R; Zhu, Z.; Yu, F. (2003). Impact of fipronil on crustacean aquatic organisms in a paddy field-fishpond ecosystem. *Bulletin of Environmental Contamination and toxicology*, 70: 746~752.

Chen, Bi-juan; Chen, Min-shan; Wu, Zhang-kuan. (1997). Study on the toxicity of fenvalerate and tetramethrin to ocean algae and shellfish. *Journal of Fishery Sciences of China*, 4(2): 51~52.

Zhang, Ai-yun; Cai, Dao-ji. (1986). Researches on ecological safety assessment for chemical pesticdes, V. A primary report and assessment on toxicity of chemical pesticides to algae. *Rural Eco-environment*, (4): 7~11.

Tang, Xue-li; Li, Yong-qi. (1999). Difference in damage of active oxygen on phaeodactylum tricornutum among three organophosphorus pesticides. *Acta Scienciae Circumstantiae*, 5 (19): 579~581.

Xie, Rong; Tang, Xue-xi; Li, Yong-qi; Wang, You. (2000). Effect of profenofos on glutathione peroxidase and on the contents of glutathione and carotinoid in two species of marine microalgae. *Journal of Ocean University of Qingdao*, 30(4) : 645~650.

Tang, Xue-xi; Li, Yong-qi; Li, Chun-yan; Dong, Bao-xian. (1995a) Biologioal Studies of the Toxicity of Organophosphorus Pesticides on Marine Microalgaes I .The Relations between the Activity of Superoxide Dismutase and Monocrotophos Tolerance in Four Marine Microalgaes. *Marine Environmental Science*, 14(2): 1~5.

Tang, Xue-xi; Li, Yong-qi; Li, Chun-yan; Dong, Bao-xian. (1995b). Biological Study of Monocrotophos Toxicity to Microalgae Ⅲ. Monocrotophos Damage to Dicrateria sp. through Free-radical. *Marine Science Bulletin*, 14(2): 29~34.

Tang, Xue-xi; Li, Yong-qi; Dong, Bao-xian. (1996). Biological Study of Toxicity of Organophosphorus Pesticides to Microalgae Ⅳ . Primitive Study of Peroxidase Inhibitors Produced by Monocrotophos Stress in Dicrateria sp. *Marine Science Bulletin*, 15(1): 27~32.

Tang, Xue-xi; Yu, Juan; Li, Yong-qi. (2001). Damage of active of oxygen to marine microalgae cells by monocrotophos. *Oceanologia Et Limnologia Sinica*, 32(2): 155~158.

Liu, Mei; Liu, Zan. (1998). Studies of Positive Growth Response to Low Concentration Organophosphorus Pesticide Dimethoate. *Journal of Jiazuo Institute of Technology*, 17(5): 396~399.

Yang, Geng; Zhang, Xiao-qiang; Qian, Zhong-ning. (2003). Research on Chronic Toxicity of MET to *Daphnia magna*. *Modern Agrochemicals*, 2(3): 22~25.

Zheng, Yong-hua; Zhao, yang; Wang Guan-yu. (1999). Studys on toxicity of meothrin to *Carassius auratus* and other several aquatic animals. *Southwest China Journal of Agricultural Sciences*, 12(3): 117~121.

Han, S.; Burnison, B. K.; Fan, G. T.. (1991). Influence of natural organic matter in water on toxicity of fenvalcrate. *J. Environ. Sci.*, 12(6) : 16~19.

Kikuchi, M.; Sasaki, Y.; Wakabayashi, M.. (2000). Screening of organophosphate insecticide pollution in water by using *Daphnia magna*. *Ecotoxicol. Environ. Safety*, 47: 239~245.

Tan, Ya-jun; Li, Shao-nan; Wu, Xiao-mao. (2004). Chronic Toxicity of Several Insecticides to *Daphnia magna*. *Chinese Journal of Pesticide Science*, 6(3): 62~66.

Xie, Rong; Tang, Xue-xi; Li, Yong-qi. (1999) Study on joint toxicity of heavy metal and organophosphorus pesticide to marine micoroalgae. *Marine Environmental Science*, 18(2): 16~20.

Qian, Yun; Zhu, Lin; Liu, Guang-liang. (2000). The joint toxic effect of pesticide mixtures on acetylcholinesterase (AchE). *Technigues and Equipment for Environmental Pollution Control*, 1(4): 27~32.

Gu, Xiao-jun; Li, Shao-nan; Fan, De-fang. (1999). The effect of fipronil and malathion mixture on topmouth gudgeon (*Pseudorasbora parva*). *Chinese Journal of Pesticide Science*, 1(1): 81~83.

Gu, Xiao-jun; Li, Shao-nan; Fan, De-fang. (2000). Effect of alternative use of fipronil and malathion on topmouth gudgeon (*Pseudorasbora parva*) AChE recovery. *Environmental Pollution and Control*, 22(4): 4~5.

Zhang, Luo-ping; Hong, Hua-sheng; Chen, Zong-tuan; Chen Wei-qi; Davide, Calamari. (1999). An Initial Environmental Risk Assessment of Pesticide Application in Xiamen Seas. *Journal of Xiamen Unversity(Natural Science)*, 38(1): 96~102.

Cai, Dao-ji; Gong, Rui-zhong; Tang, Guo-cai; Chen, Liang-yuan. (1997). Safety Evaluation of Deltamethrin on Aquatic Organisms in a Ricefield. *Research of Environmental Sciences*, 10(3): 30~35.

Shan, Zheng-jun; Wang, Lian-sheng; Cai, Dao-ji; Gong, Rui-zhong; Zhu, Zhong-lin. (2001). Impacts of Isazofos on aquatic organisms in paddy fish pond ecosystem. *Journal of Nanjing Agricultural University*, 24(3): 93~96.

Shan, Zheng-jun; Wang, Lian-sheng; Cai, Dao-ji; Gong, Rui-zhong; Zhu, Zhong-lin; Yu, Fei. (2002) Impacts of Fipronil on Aquatic Organisms in Paddy-fish Pond Ecosystem. Scientia Agricultura Sinica, 35(8): 949~952.

Pan, Dao-yi; Liang, Xue-Ming. (1996) The susceptibility of marsh frog (tadpole) and spiders to pesticides and classification of acute toxicity. Acta Zoological Sinica, 42(2):154~160.

Pan, Dao-yi; Ruo, Rong; Tie, Bai-qing; Li, Jian-hong. (1989) Toxicity tests of funigicide to the tadpoles of bog frog (*Rana Limnocharis*). Journal of Hunan Agricultural College, 15(4): 46~50.

Pan, Dao-yi. (1990) Toxicity of herbicides to the tadpoles of bog frog (*Rana Limnocharis*). Chinese Journal of Zoology, 25(1): 32~34.

Nie, Hai-yan; Huang, Yu-yao. (2004) Prediction of the Effect of DMAH on the Structure and Function of Model Pond Ecosystem. Journal of Central South Forestry University, 24(2):18~22.

Xu,Fu-liu; Dawson, R. W.; Tao, Shu; Li, Ben-gang; Xu, Jun-cao. (2002). System-level responses of lake ecosystems to chemical stresses using exergy and structural exergy as ecological indicators. Chemosphere, 46(2): 173~185.

In: Aquatic Toxicology Research Focus
Editor: Elias P. Svensson, pp. 99-118
ISBN: 978-1-60456-192-0

Chapter 5

ECOTOXICOLOGY OF HEAVY METALS IN MARINE SPONGES: DIFFERENT AND CONTRASTING EFFECTS OF HEAVY METALS ON DIFFERENT BIOLOGICAL LEVELS

Emma Cebrian[1]
Centre d'Estudis Avançats de Blanes. C/ Accés a la Cala Sant Francesc 14.
17300 Blanes. Girona. Spain

ABSTRACT

Toxic effects of heavy metals are a widespread concern in the modern toxicological studies. However, most studies dealing with the effects of heavy metals on invertebrates have addressed a few aspects of the organisms biology, and rarely attempt to analyse heavy metal effects at different levels of biological organization. The current trend in ecotoxicological studies proposes a use of a wide range of approaches involving several methods, each one focusing on different level of biological organization, to acquire a better understanding of noxius effects of heavy metals on marine invertebrate populations. The same metals may have contrasting effects on marine invertebrates depending on the biological level considered. As an example, here we present a review of several studies addressed to assess the ecological effects of sublethal concentrations of heavy metals on marine sponges. Responses at a physiological, molecular and early stages development levels differ according to the species and metal considered.

Sponges submitted to sublethal concentrations of heavy metals experience several behavioural and physiological responses such as changes in shape, metal accumulation, decreases in growth rates, reproduction output and filtration rates. Species differences in adult tolerance to heavy metals may be explained by differences in HSP expression. Induction of HSP production may allow some sponge species to endure heavy metal contamination. Studies on the effects of heavy metals on early development stages of sponges have shown contrasting results. While chronic expositions to heavy metals were harmful against adults, short expositions to similar concentrations of heavy metals resulted innocuous for larvae and juveniles of the same species. Thus, in order to acquire

1 E-mail: emma@ceab.csic.es.

a general view of the effects of heavy metal on sponges, perhaps we should take into consideration all punctual responses given at different organization levels from organisms to molecules in order to avoid misinterpretation of the results.

INTRODUCTION

Toxic effects of heavy metals are a widespread concern in the modern toxicological studies. In general, heavy metals produce their toxicity by forming complexes or "ligands" with organic compounds. Thus, modified biological molecules lose their ability to function properly and result in a malfunction or death of the affected cells.

Up to now, the literature is replete with examples of metal quantification in water [1-3] sediments [4-6] and biota [7-10]. Their limitation is, however, that they do not provide information regarding ecological/biological effects of a particular metal concentration. In this sense, several authors underline the need for information on the relationships between bioaccumulation and effects [9, 11]. Although in the literature there are several examples on metal accumulation in the aquatic environments, it is extremely difficult to accurately assess the toxicological significance of metals present in a given site, and to attribute toxic effects to these metals [11].

Up to now, the several examples documenting severe toxic effects of heavy metals on marine invertebrates, mainly concern aspects such as death [e.g. 12-13], growth inhibition [e.g. 14-16] and reproduction decrease [e.g. 17-18], among others. However, most of these studies have studied metal toxicity in the laboratory [e.g. 19-22]. Unfortunately, there is a lack of ecological relevance of many standard ecotoxicological test, besides the extrapolation from laboratory findings to real world situations are often impractical [23-24].

Ecological effects of heavy metals on organisms, populations and ecosystems may arise in different ways and magnitudes. Some examples are reported on the occurrence of acute heavy metal pollution on sessile species that drastically change the composition and spatial distribution of sublittoral communities [e.g. 25-26]. However, most often, heavy metals are released to marine coastal ecosystems in low concentrations, which produce sublethal effects in organisms with repercussion in the communities at a long-term. Among the effects of heavy metals on marine environments, population-level effects are perhaps the most neglected area in ecotoxicological research [27]. Effects of heavy metals on populations may involve loss of individuals and can be assessed by studying their effects on different levels of biological organization, from molecules to populations: e.g. individual growth rates, reduced fecundity and longevity, disturbance of endogenous biological rhythms and molecular and cellular defences [28-30].

To investigate contaminant exposure and effects of heavy metals on living organisms, ecologist have realised that biomarkers are powerful tools [27, 31]. The simultaneous use of several biomarkers that reflect effects and toxic exposure at various levels of biological organization will provide information on a greater number of consequences of a given heavy metal concentration on populations, and thus, ecosystems.

Physiological and behavioural markers provide information on the effects (injuries) of heavy metals at organism and population levels and, as a consequence, at the community level (Table 1). Growth, reproduction output, recruitment success, filtration rates, morphological changes, and structural elements, are among others, the most commonly measured biomarkers that indicate physiological responses at the organism level, which probably make itself felt on populations and communities; however changes and effects on those biomarkers are detectable at a long term (days, months and years; Table 1). On the other hand, most molecular biomarkers respond faster to toxic exposure than physiological ones (Table 1).

Table 1.1. Some examples of biologic markers employed in ecotoxicological studies (modified from Evenden and Depledge, 1997)

Level of organization	Type effect	Biological marker	Type of marker	Time scale
Biological molecules	Molecular responses	Genes expression: cell regulation and defense/repair proteins: (e. g. HSPs and metal-binding proteins)	Markers of exposure	Seconds
Cell	Tissue and cellular effects	Cell aggregation and motility	Markers of effects	Minutes
Individual	Physiological effects and disease	Reduction in growth rates Changes in reproductive output Respiration, Clearance rates		Hours
	Impaired reproductive capability Death		Markers of ecological change	Days
Population	Decline in reproduction Reduced *sensus* numbers	Deformities Death Population decline Change of population density		Months
	Population decline			
Community	Possible extinction	DNA analyses	Markers of evolutionary change	
Ecosystem	Reduction in biodiversity	Changes in ecological descriptors (e.g. biodiversity, richness, similarity.. indices)		Years

As each type of biomarker has its own pros and cons, integrated studies involving biochemical, physiological and behavioural biomarkers in order to assess the ecological effects of heavy metals on populations and ecosystems are required. However, most of the studies dealing with heavy metal toxicity on marine invertebrates have addressed a few aspects of the organisms biology and have used few biomarkers, being in this way, very difficult to provide information regarding on the ecological/biological effects of a particular metal concentration. Furthermore, generating general assumptions from punctual responses would lead us to erroneous conclusions, since responses to heavy metals vary depending on several factors, which should be considered for a real understanding of the global effects of a fixed toxicant in the ecosystem.

Sponges appear to be a suitable group for this kind of studies because besides to be a very important group in marine coastal environments [32], they process large amounts of water [33-35], show a wide distribution and year-round availability, have a high capacity for accumulating heavy metals [36-38] and furthermore, the extensive knowledge of the biology and physiology at different levels of biological organization (molecular, larval, settlers, juveniles, organisms) necessary to achieve this sort of studies is available.

Hence, as an example of the need of studies dealing with the effects at different levels of biological organization, here we show the contrasting and in some cases controversial effects of heavy metals found on sponges, depending on the biomarker or the biological level considered. This review shows that only the integration of different studies concerning several levels of biological organization, heavy metal concentration and exposure and species, may lead us to a correct comprehension of heavy metal toxicity.

Ecological Effects of Heavy Metal Pollution on Sponge Physiology and Populations

The effect of low levels of copper contamination on the sponges physiology and behaviour has been examined in two Mediterranean sponges (*Crambe crambe* and *Chondrosia reniformis*) [39, 40]. Individuals of both species were transplanted to the same copper polluted site and accumulation, shape indices, growth and survival were monitored in both species during four months. Both species behaved differently in front of the same pollution type. Specimens of *Crambe crambe* transplanted to the polluted site were able to accumulate copper, while specimens of *Chondrosia reniformis* did not experienced any heavy metal accumulation (Figure 1). On the other hand, area and the perimeter/area relation, which indicate the number of fissions experienced by an individual in consequence of stress conditions [41], also varied in *Crambe crambe* specimens transplanted to the polluted site. In contrast, *Chondrosia reniformis* specimens did not experienced any change neither in area nor perimeter/area relation (Figure 2). The irregularity and fission increment observed in *C. crambe* specimens transplanted to the polluted site, probably are indicating stress conditions due to copper pollution. Likewise, growth decreased in *C. crambe* when was submitted to copper pollution, while specimens of *C. reniformis* grew similarly in copper contaminated and control conditions (figure 3).

Accordingly to the cited experiment, we could suggest that *C. crambe* is more sensible than *Chondrosia* since it accumulated copper, suffered growth decrease and fissions in face to copper pollution. In contrast, *Chondrosia*, at first sight, seems to be resistant to this environmental pollution, as it did not experience any physiological and behavioural change. Thus, a study, which only would take into account the cited variables, might conclude that among both, *Chondrosia* is the most resistant specie to copper pollution. Anyway, concerning to survival of both species a sudden mortality was observed during the last month of the experiment in *C. reniformis* specimens, while survival was not affected in *Crambe* (figure 4). Therefore, the lack of responses in *Chondrosia* did not meant a lack of effect, but variables analysed were suitable to detect heavy metal effect on *Crambe* but were unfit for *Chondrosia*.

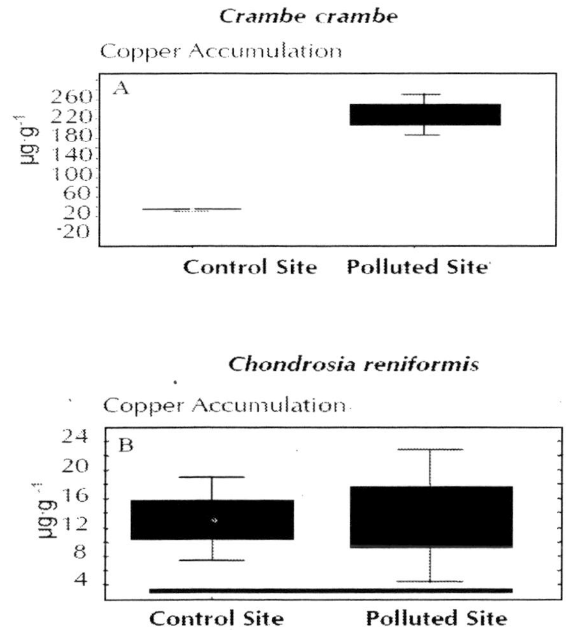

Figure 1. Mean concentration of copper within the sponges (A) Crambe crambe and (B) *Chondrosia reniformis*. C: control site; P: Polluted site. Boxes represent standard errors; Vertical bars are standard deviations. Mean concentrations, which proved not significantly different in Tukey test were joined by horizontal lines. Modified from Cebrian et al. 2003 and 2006.

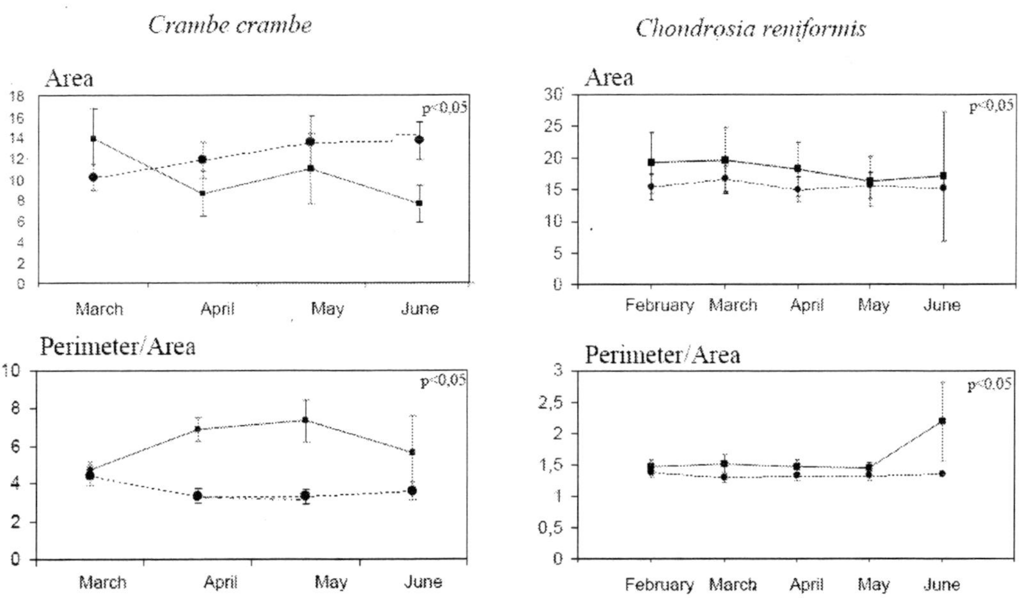

Figure 2. Time course of the mean sponge area and perimeter/area ratio of the sponges (*Crambe crambe* and *Chondrosia reniformis*) of the control site (dashed line) and of the polluted site (solid line). Vertical bars are standard errors. Modified from Cebrian et al. 2003 and 2006.

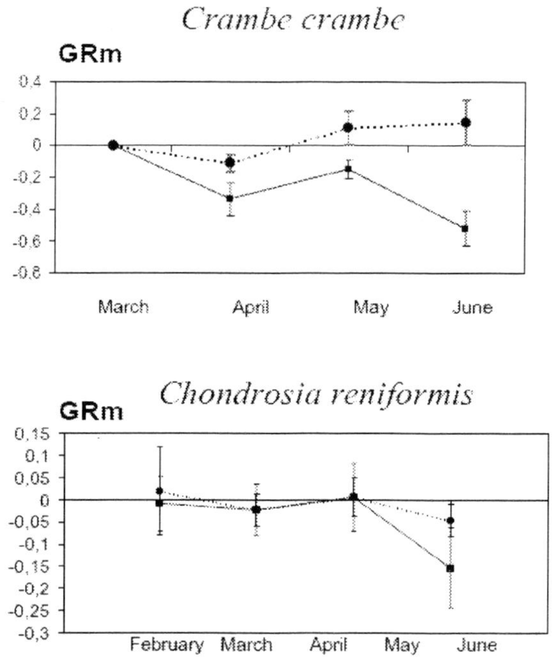

Figure 3. Time course of the monthly growth rate of Crambe crambe and Chondrosia reniformis of the control (dashed line) and polluted site (solid line). Vertical bars are standard errors. Modified from Cebrian et al. 2003 and 2006.

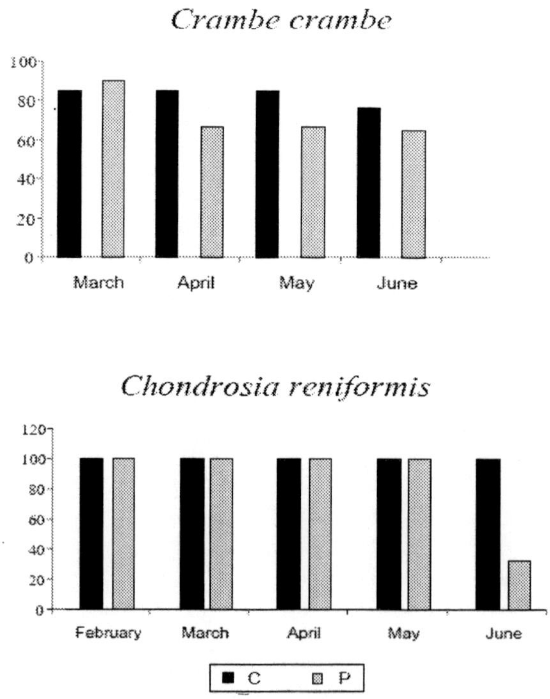

Figure 4. Percentage of survivors among sponges of the polluted (P) and control (C) site. Modified from Cebrian et al. 2003 and 2006.

In fact, clearance rates were significantly lower in *Chondrosia* transplanted to the copper polluted site compared with those of the unpolluted site [40]. It seemed, as it has been suggested for other species and other heavy metals [37], that copper reduced filtration rates in *Chondrosia*, leading to harmful consequences detectable at a longer term, since a continued reduction in water pumping may result in cell starvation, which has been reported to induce cell apoptosis [42], what at last would explain the posterior mortality observed in *Chondrosia* specimens. Furthermore, the decrease in pumping rates may also explain the lack of copper accumulation in the specimens of *Chondrosia* submitted to copper pollution. These results show that even we use a wide range of biomarkers (accumulation, growth and fission) in both sponges to detect the effects of heavy metals, the conclusions obtained might be erroneous if all aspects of the sponge physiology are not taken into account, as for example, if we neglect *Chondrosia* filtration rates and mortality at a longer-term (four months).

In resume, these studies have shown that even in the same group, effects and responses to heavy metals can be very different among species, and that even the most commonly used physiological variables may not provide the real effects of heavy metals on organisms. In this way, some authors [43-47] have suggested that the study of biochemical and molecular markers might provide rapid and useful information, since one of the most important features of molecular/cellular biomarkers is that they have the potential to anticipate changes at higher levels of biological organisation, i.e.population, community and ecosystem [48].

Ecological Effects of Heavy Metal Pollution on Sponges at a Molecular Level

In this way, at a molecular level the suitability of stress proteins (sometimes referred as heat-shock proteins) as biomarkers of adverse effects by metals has been demonstrated for other invertebrates [49]. In sponges, HSP have proved to be induced by non-ionic organic fraction from a polluted river [50], some PCBs [51] and cadmium [52]. Thus, HSP were used as a molecular marker to achieve the effect of low levels of copper contamination in *Crambe crambe* and *Chondrosia reniformis* [40, 53]. In these studies, HSP induction expression was analysed in the specimens of both species transplanted to a copper polluted and to an unpolluted site. Furthermore, laboratory experiments in which variables could be controlled and their effects quantified were run in parallel to confirm field results. As a result, *Crambe crambe* specimens submitted to moderate concentrations of copper both *in situ* and at the laboratory induced the production of two proteins of different molecular weight, HSP 54 and HSP72. HSP 54 accumulated in the long term (sponges living for more than 4 years at the polluted site; pers. obs.) (Figure 5A), and it seemed that copper tolerance was enhanced in *C. crambe* as long as the expression of HSP54 was high. In contrast, HSP72 responded faster but was less persistent, being mainly induced at short term against copper pollution (it was induced only in specimens transplanted to the polluted site during 4 months) (Figure 5B). In the laboratory when *C. crambe* was submitted to similar concentrations of copper pollution (aprox 30μg.l^{-1}) during one week, no induction of HSP54 was observed (Figure 5C), in contrast, HSP72 expressed similarly than in the field (Figure 5 D). However, under higher concentrations (100μg-l^{-1}) induction of both proteins was inhibited (Figure 5C,D).

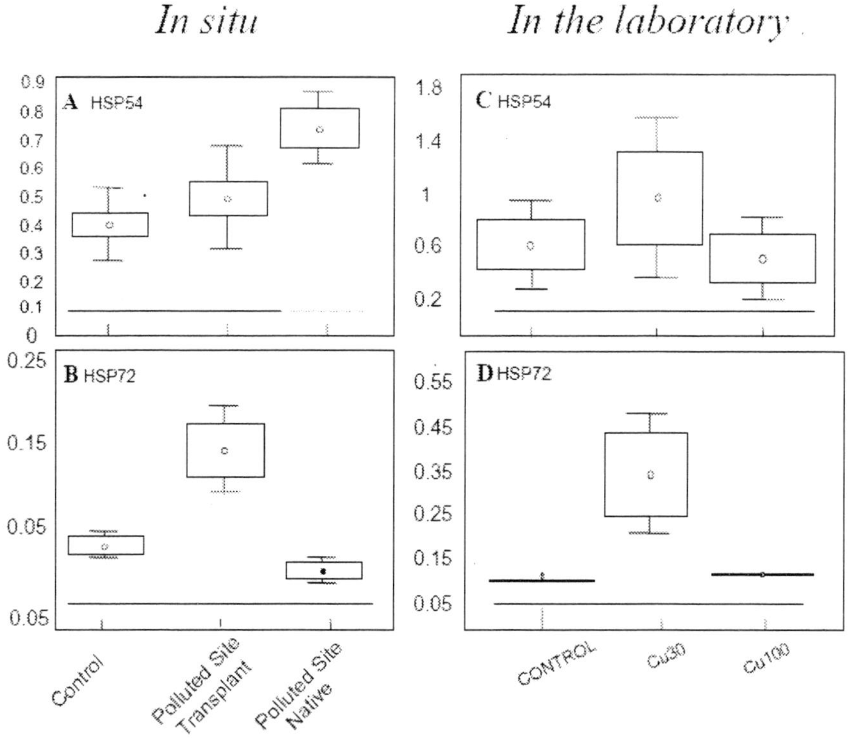

Figure 5. Densiometer readings of the HSP 54 and HSP72 bands in *Crambe crambe* specimens from the field and laboratory experiments. Protein amounts which proved not significantly different in the Tukey test were joined by horizontal lines. The dash line underlines the treatment that significantly differs (Tukey test) from those underlined by a continous line. Boxes and vertical bars are standard errors and standar deviations, respectively. Modified from Agell et al. 2001.

On the other hand, when specimens of *Chondrosia reniformis* were submitted to the same conditions, although it expressed a band of 60 kDa, no induction of HSP was observed among specimens of the polluted and the non-polluted site. (Figure 6 A, B). Correlation has been reported between the amount of stress proteins expressed and the ability to survive subsequent, more severe stress, which would otherwise be lethal to the organism [45, 54]. Thus, it seems that at a molecular level, *C. crambe* is more protected against heavy metal damage thanks to its ability to induce defence mechanisms when submitted to copper pollution. On the other hand, if studies on the expression of HSP on *Crambe* were only performed in the laboratory and with copper concentrations higher than 100 $\mu g{\cdot}l^{-1}$, as many studies on HSP expression in invertebrates [55-57] the results would not show repair mechanisms against copper pollution, when actually the mechanisms exist but probably are inhibited over a threshold concentration due to a metabolic damage as it has also been ascribed for other organisms [55, 58]. Similarly, the lack of molecular response in *C. reniformis* might seem at first sight, that heavy metals have no effect on this sponge species, however the mortality observed on *C. reniformis* after prolonged exposition to copper pollution, suggests that the lack of response is due to a strong metabolic damage.

Again, these results taken together show that the use of a unique biomarker might drive us to incorrect conclusions, and that it is necessary to go through as much as possible biomarkers to ascertain the real effects of heavy metals on individuals and populations.

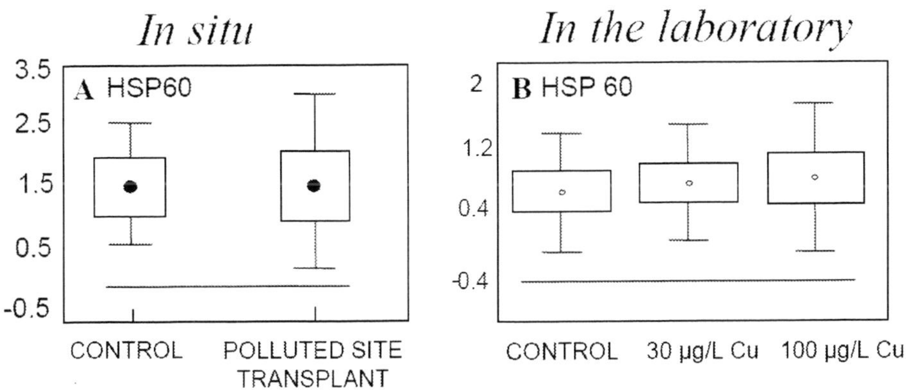

Figure 6. Densiometer readings of the HSP 60 in *Chondrosia reniformis* specimens from the field and laboratory experiments. Protein amounts, which proved not significantly different in the Tukey-test were joined by horizontal lines. Boxes and vertical bars are standard errors and standar deviations, respectively. Modified from Cebrian et al. 2003.

In this way, differences in tolerance to heavy metal pollution observed between *C. crambe* and *C. reniformis* by means of the physiological biomarkers, might be explained by differences in HSP expression. In fact, *C. crambe* induces HSP production and actually it can live in moderately copper polluted areas, while *C. reniformis* do no enhance HSP production under the same conditions and neither inhabits these moderately polluted areas [59].

ECOLOGICAL EFFECTS OF HEAVY METAL POLLUTION ON EARLY STAGES OF DEVELOPMENT

Sublethal effects of heavy metals may have drastic repercussions at an ecological level when they alter biological processes of the organisms that indirectly affect to successive populations [60]. For instance, a pollutant may kill a half of the individuals of a species population with little or no ecological significance, whereas a pollutant that does not kill organisms but retard development may have a considerably higher ecological impact [29]. Therefore, changes that affect populations overall reproductive capability (premature death, fecundity and viability of offspring, ect.) are of great influence on effects at a higher levels of biological organisation, and it is therefore to consider the study of this phenomena as an important ecotoxicological endeavour [61].

In many invertebrates early stages are more sensitive to pollutants than adults [60, 62-63] and similarly, we expected that low level of copper pollution which were not necessary lethal for adult sponges may have greater effects on larvae and juveniles largely determining a subtle decline and even disappearance of sponge populations in polluted areas. In this way [64] studied the effects of low levels of heavy metal pollution on sponge larvae and juveniles of *Crambe crambe*. Therefore larvae of *Crambe crambe* were submitted to copper (30 $\mu g \cdot l^{-1}$) and cadmium (5 $\mu g \cdot g^{-1}$) during one week in the laboratory. Selection of heavy metal concentrations was done according the approximate concentrations that can be found in marine polluted zone areas [2, 59]. As a result, copper and cadmium did not effect larval settlement on *Crambe crambe* (figure 7 A), what may seem at first glance surprising, as in general, heavy metals have been reported to negatively affect larvae [65-66] and metamorphosis [67] of a great number of invertebrates.

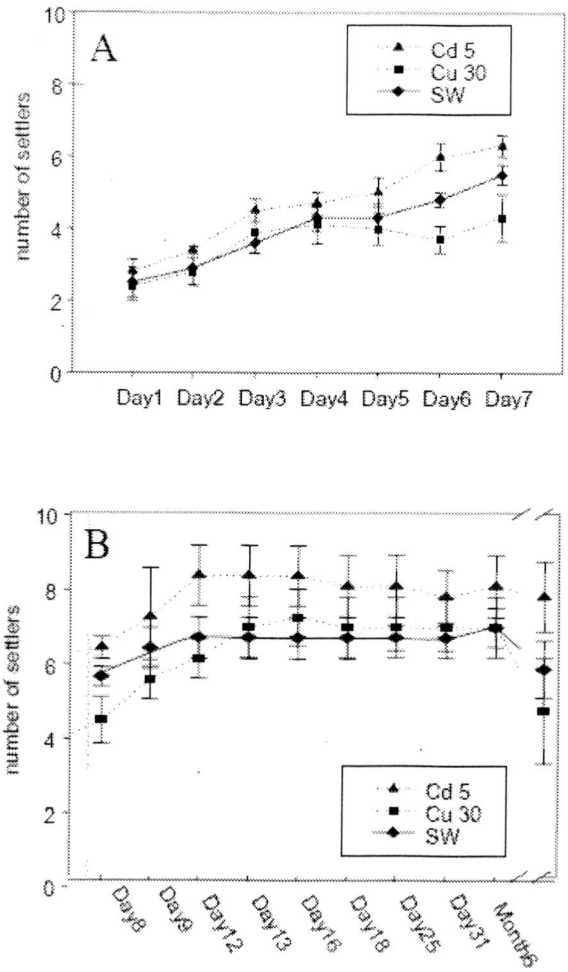

Figure 7. Settlement of *Crambe crambe* larvae with copper (30 $\mu g \cdot l^{-1}$) and cadmium (5 $\mu g \cdot l^{-1}$) as a function of time (A); and survival of *Crambe crambe* juveniles previously incubated under Cu and Cd (concentrations as in figure 7A) during 6 months. Vertical bars are standard errors. Modified from Cebrian et al. 2007.

Afterwards, [64] analysed the posterior capacity of survival in clean seawater of the settlers previously incubated with heavy metals, in order to know if short pulses of heavy metals although do not having effects on settlement, may have posterior effects on the juvenile survival and thus, success, of sponge populations. Accordingly, the authors did not find significant differences in survival between settlers coming from larvae previously incubated with low concentrations of copper and cadmium and those maintained in SW control (figure 7B) and thus, we can assume that low concentrations of copper and cadmium did not exert any effect on the consecutive survival and growth of the sponge juveniles. Similar experiments were performed *in situ* in order to corroborate results obtained in the laboratory. When juveniles of *Crambe crambe* previously settled in the laboratory, were transplanted *in situ* to a control site and to a copper polluted site with similar copper concentrations than those assayed at the laboratory, a comparable pattern of settler mortality was observed at both, the polluted and control sites (figure 8) [39]. Although juvenile mortality was significantly higher *in situ* than in the laboratory, probably to a major number of uncontrolled variables affecting settlers, the important conclusion was that *in situ* as well as at the laboratory, copper did not exert any negative effect on *C. crambe* larvae, settlers and juveniles. Therefore, these former results might lead as to think that except the observed sublethal effects of copper pollution on *C. crambe* adult physiology, as are copper accumulation, growth decrease and a fission increment, no major repercussions of low levels of copper pollution could be expected in *Crambe crambe* populations, thanks to the induction of repair mechanisms (HSP). Conversely, when the percentage of individuals undergoing reproduction at the copper polluted and at the unpolluted site was assessed, considerable differences on reproduction were observed between individuals of both sites. The percentage of sponges harbouring embryos of individuals naturally occurring at the polluted site was significantly lower (11,7%) than in those of the unpolluted site (82%) (Figure 9) [39]. Thus, as it has been demonstrated for other invertebrates [68], it seems that copper inhibits embryo development in *C. crambe*. Sponges mainly rely on their larval settlement for their maintenance [69], thus, although moderate concentrations of copper did not exhibit any effect on larval settlement, its effects on embryo development would indirectly affect larval settlement and thus, sponge population success. Again, the study of the effects of heavy metals on a single process, as for example settlement success, would avoid us to realise of the real ecological effect of a given heavy metal burden on sponge populations.

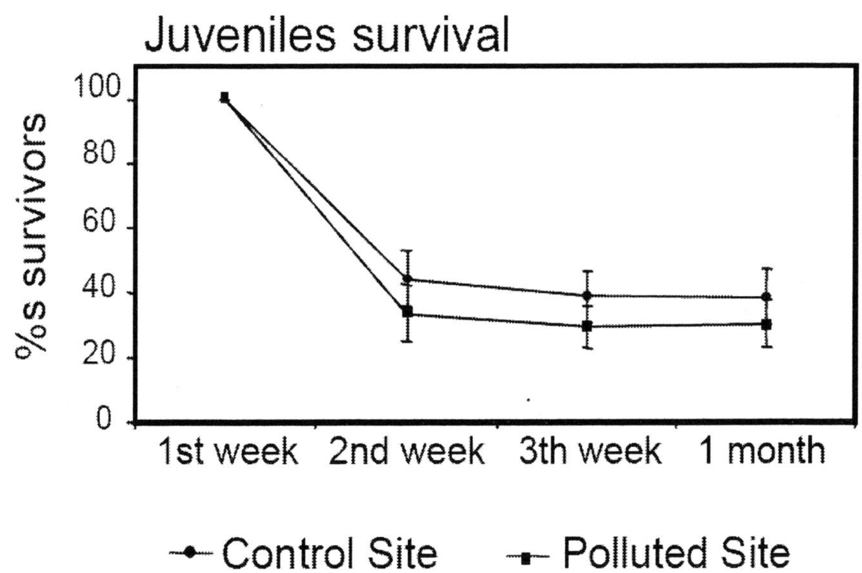

Figure 8. Juvenile survival (%) at the control site (solid symbol) and the polluted (empty symbol) site. Vertical bars are standard errors. Modified from Cebrian et al. 2003.

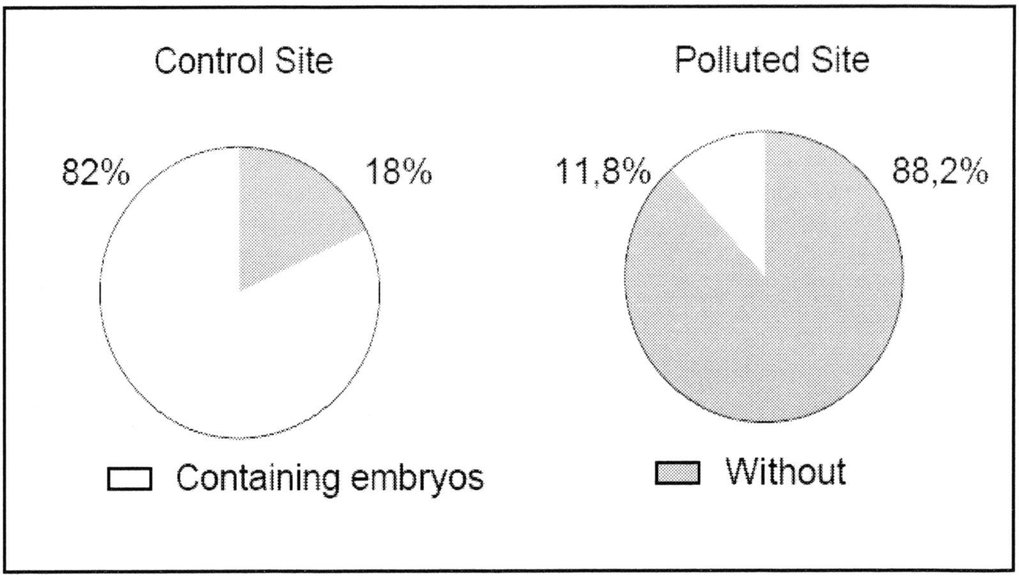

Figure 9. Percentage of sponges harbouring embryos in the control and polluted site. Modified from Cebrian et al. 2003.

Conclusion

To assess the ecological effects of pollutants, and concretely heavy metals, at a population and community levels is of great interest, but at the same time it has high difficulties because of the great number of environmental stressors that must be monitored, the high cost and long duration of ecosystem-scale test, the problem of assessing synergisms, ect [70]. Is because of that, the prediction of ecosystem stress must rely heavily on bioassays and monitoring at simpler scales [70]. So far, among biomarkers at simpler scales, there has been a great effort directed to study biomarkers of exposure, such is accumulation, to know polluted areas. In fact, the literature is replete with examples of metal quantification in an array of marine organisms collected from numerous locations [10, 71-73]. However in general, accumulation and toxicity are not reported together and then, these studies do not provide information regarding on the ecological/biological significance of a particular metal burden. Furthermore, differences in accumulation among species [74-76], demonstrate the difficulty in predicting toxic effects from chemical measurements alone, and the danger of extrapolating results from one species to another. For instance, in the present chapter and as an example, we have seen that heavy metal accumulation by itself can not give us an idea of, first, the pollution degree of a particular site, and second, the effects of heavy metals on sponges. Otherwise, according to the metal accumulation observed in *Chondrosia reniformis*, we would conclude that it was living in a non-polluted site, and second that copper did not affect *Chondrosia reniformis*, when actually was causing a deep metabolic damage with fatal consequences at a long-term.

Aside from accumulation, most of the studies dealing with the heavy metal effects on marine ecosystems use simple approaches concerning to simple levels of biological organization, which afterwards will provide hints on the effects at more complex organization levels. Going through the available bibliography, except few studies [47, 77] we can realise that in general, studies on the effects of heavy metals on marine invertebrates have addressed a few aspects of the organisms biology, taking into account one [e. g. 19, 66, 78-81] or two [e. g. 19, 82], at most three biomarkers [83]. However, as it has pointed out the present chapter, these previous and punctual studies may provide anecdotic results if they neglect the different responses experienced at different levels of biological organization. For instance, here we have seen that even within the same species, responses against copper pollution differ depending on the biomarker considered. While no effects were observed on *Crambe crambe* settlers, a decrease in reproduction effort, and growth was observed in *C. crambe* adults. Other example is that in *Chondrosia reniformis* copper did not affect induction of HSP at short term, neither growth decrease nor fission, but after 4 months exposition, mortality significantly increased in the copper polluted site.

Therefore, as it has been pointed out by several authors [27-28, 30], effects of heavy metals at high levels of biological organisation should be tackled by means of multidisciplinary studies considering a great range of biomarkers (here growth, fissions, survival, HSP, settlement, reproduction), because if we would only considered some of them, for example induction of HSP, accumulation, fissions and growth in *C. reniformis*, which number is considerably higher than the considered in most of ecotoxicological studies, we would not attain the real ecological significance of the observed effects.

Furthermore, it is also widely recognised that heavy metals effects and so, biological responses, considerably vary among species [47, 59, 75-76, 84], what show the danger of extrapolating results from one species to another. But in contrast, the literature is full of examples of the effects of heavy metal considering different biomarkers besides on different species such are algae [55], fishes [e.g. 85-86], sponges [36, 38, 84], tunicates [26, 77, 87], cnidarians [66-67], briozoans [88], polychaetes [89] among others, what complicates comparisons and thus a global comprehension on the real effects of toxicants on the ecosystems. For example, the present chapter has shown that even two close species with similar behaviour, respond in a very different way to the same pollution source, and thus, the observed responses and effects for one species are not valid neither comparable for the other. *C. crambe* induced HSP, decreased growth and experienced fissions, while *C. reniformis* did not, in face to the same copper pollution.

Nevertheless, up to now, mussels are among marine benthic organisms, the only group studied considering the same or very similar species at several levels of biological organization from molecules [e. g 90-92] cellular [e.g. 93, 94] and physiological [e.g. 95-97]. It is not surprising, since a common motive for selecting particular species for ecotoxicology studies of heavy metals appears to be the need to investigate the potential for heavy metal transfer to humans through the ingestion of seafood [98]. Anyway, it must be remembered that the approaches as mentioned above can predict biological effects on the test species (i.e. mussels), but not necessary on the ecosystem as a whole, and maybe the effects observed are not comparable with the effects on other species. Thus, studies on the effects of heavy metals on as much species as possible, but on the other hand, simultaneously studying each species at different levels of biological organization, are a fundamental request to be able to integrate the different results necessary to acquire the real ecological significance of heavy metal effects on whole ecosystem.

In this sense, in this chapter we have tried to document the effects of moderate copper pollution on sponge population by means of studying its effects in several simple scales in the field and in the laboratory in two different species. As a result we can conclude that no single approach is satisfactory, and this chapter emphasises the value of an integrated assessment approach involving several methods and each focusing at several levels of biological organisation, in order to avoid misinterpretation of the results and to acquire a general view of the effects of heavy metal populations and organisms. In contrast, the main literature available shows that most of the studies focus in only few biomarkers, studied in very different species, which cannot provide the general frame necessary to understand the real effects of heavy metal pollution on marine ecosystems. Thus, although we know about the existence of heavy metal polluted environments for a long time, we are still unacquainted the real effects of those determined metal burden.

Acknowledgements

I gratefully thank to all the people who has unselfishly helped in all the fieldwork and laboratory analyses necessary to perform all the work here presented. I also acknowledge to all anonymous referees that have improved our manuscripts with their useful comments.

References

[1] Cossa, D., Elbaz-Poulichet, F., Gnassia-Barelli, M., Roméo, M. (1993). Le plomb en milieu marin. Biogéochimie et écotoxicologie. Repères océan 3, IFREMER.

[2] Cotté-Krief, M.H., Thomas, A., Martin, J.M. (2002). Trace metal (Cd, Cu, Ni and Pb) cycling in the upper water column near the shelf edge of the European continental margin (Celtic Sea). *Marine Chemistry* 79: 1-26. [10] Rainbow PS (1995). Biomonitoring of heavy metal availability in the marine environment. *Marine Pollution Bulletin* 31: 183-192.

[3] Accornero, A., Manfra, L., Salluzzo, A., Modestia, F. (2004). Trace metal pollution in surface marine waters: nearshore concentrations along Apulia and Albania. *Chemistry and Ecology* 20 (1): 195-203.

[4] Bryan, G.W., Langston, W.J. (1992). Bioavailability, accumulation and effects of heavy metals in sediments with special reference to United Kingdom estuaries: a review. *Environmetal Pollution* 76: 89-131.

[5] Palanques A, Díaz J.I., Farrán, M. (1995). Contamination of heavy metals in the suspended and surface sediment of the Gulf of Cadiz (Spain): the role of sources, currents, pathways and sinks. *Oceanologica A*cta 18: 469-477.

[6] Puig, P., Palanques, A., Sanchez-Cabeza, J.A., Masqué, P. (1999). Heavy metals in particulate matter and sediments in the southern Barcelona sedimentation system (North-western Mediterranean). Marine Chemistry 63(3-4): 311-329.

[7] Phillips, D.J.H., Rainbow, P.S. (1993). Biomonitoring of trace aquatic contaminants. New York: Elsevier Press.

[8] Rainbow, P.S., Phillips, D.J., Depledge, M.H. (1990). The significance of trace metal concentrations in marine invertebrates. A need for laboratory investigation of accumulation strategies. *Marine Pollution Bulletin* 21(7): 321-324.
[9] Rainbow, P.S. (1995). Biomonitoring of the heavy metal bioavailability in the marine environment. *Marine Pollution Bulletin* 31: 183192.
[10] Rainbow, P.S., Phillips, D.J.H. (1993). Cosmopolitan biomonitors of trace metals. *Marine Pollution Bulletin* 26: 293-601.
[11] Borgmann, U. (2000). Methods for assessing the toxicological significance of metals in aquatic ecosystems bio-accumulation-toxicity relationships, water concentrations and sediment spiking approaches. *Aquatic Ecosystem Health and Management* 3: 277-289.
[12] Arnott, G.H., Ahsanullah, M. (1979). Acute Toxicity of Copper, Cadmium and Zinc to Three Species of Marine Copepod. *Australian Journal of Marine and Freshwater Research* 30, 63–71.
[13] Casillas, E., Misitano, D.A., Plesha, P.D., Weber, D.D., Haley, C.R., Demuth, S. Schiewe, M.H., Sin-Lam Chan, Varanasi, U. (1989). Sublethal And Lethal Effects In Two Marine Organisms, A Juvenile Echinoderm And A Larval Fish, Exposed To Contaminated Sediments. *OCEANS '89. Proceedings Publication* 2: 402-407
[14] Tewari, A., Joshi, H.V., Raghunathan, C., Kumar, V.G.S., Khambhaty, Y. (2001). Effect of heavy metal pollution on growth, carotenoid content and bacterial flora in the gut of *Perna viridis* (L.) in in situ condition. *Current Science* 81(7): 819-828.
[15] David, C.P. (2003). Heavy metal concentrations in growth bands of corals: a record of mine tailings input through time (Marinduque Island, Philippines). *Marine Pollution Bulletin* 46(2): 187-196.
[16] Yang, S., Wu, Q. (2003). Effect of Cd on growth and physiological characteristics of *Aegiceras corniculatum* seedlings. *Marine Environmental Science* 22(1): 38-42.
[17] Bhattacharya, S.S., Vaidya, A.A. (1999). Effect of copper on reproduction of the amphipod *Parhyale fascigar* (Stębbing). The Fourth Indian Fisheries Forum, Proceedings. pp. 117-120.
[18] Daka, E.R., Hawkins, S.J. (2002) Reproductive effects of heavy metals on the rough periwinkle *Littorina saxatilis* (*Mollusca: Prosobranchia*). *Indian Journal of Marine Sciences* 31(3): 218-224.
[19] Brown, B., Ahsnullah, M. (1971). Effect of Heavy Metals on Mortality and Growth. *Marine Polution Bulletin* 2: 182-185.
[20] Depledge, M.H. (1984). Disruption of endogenous rhythms in Carcinus maenas (L.) following exposure to mercury pollution. *Comparative Biochemistry and Physiology Part A* 78: 375-379.
[21] Rainbow, P.S., Wang, W.X. (2001). Comparative assimilation of Cr, Se and Zn by the barnacle *Elminius modestus* from phytoplankton and zooplankton diets. *Marine Ecology Progress Series* 218: 239-248.
[22] Webster, N.I., Webb, R.I., Ridd, M.J., Hill, R.T., Negri, A.P. (2001). The effects of copper on the microbial community of a coral reef sponge. *Environmental Microbiology* 3(1): 19-31.
[23] Cairns, J., Pratt, JR. (1989). The scientific basis of bioassays. *Hydrobiologia* 188: 5-20.
[24] Giddings, J.M. (1980). Field experiments. In F.S. Sanders, S.M. Adams, L.W. Barnthouse, J.M. Giddings, E.E. Huber, K.D. Kumar, D. Lee, B. Murphy, G.W. Suter, and W. Van Winkle, (eds.), Strategies for Ecological Effects Assessment at DOE Energy

Activity Sites, pp. 315-331. Environmental Sciences Division Publication 1639. ORNL/TM-6783. Oak Ridge, National Laboratory, Oak Ridge, TN.

[25] Diaz-Castaneda, V., Richard, A., Frontier, S. (1989). Preliminary results on colonization, recovery and succession in a polluted area of the southern North Sea (Dunkerque's Harbour, France). *Scientia Marina* 53 (2-3): 214-224.

[26] De Caralt, S., López-Legentil, S., Tarjuelo, I, Uriz, M.J., Turon, X. (2002). Contrasting biological traits of *Clavelina lepadiformis* (Ascidiacea) populations from inside and outside harbours in the western Mediterranean. *Marine Ecology Progress Series* 244: 125-137.

[27] Depledge, M.H., Hopkin, S.P. (1993). Methods to Assess Effects on Brackish, Estuarine, and Near-Coastal Water Organisms. In. SCOPE 53. IPCS JOINT Activity 23. SGOMSEC 10. Methods to Assess the Effects of Chemicals On Ecosystems. Chichester : J. Wiley and Sons, pp 125-149.

[28] Gray, J.S. (1979). Pollution-induced changes in populations. Philosophical Transactions of the Royal Society 286: 545-561.

[29] Moriarty, F. (1983) Ecotoxicology: The Study of Pollutants in Ecosystems. Academic Press, London, 233 pp.

[30] Depledge, M.H. (1984). Disruption of endogenous rhythms in Carcinus maeas (L.) following exposure to mercury pollution. *Comparative Biochemistry and Physiology Part A* 78: 375-379.

[31] Depledge, M.H., Amaral-Mendes, J., Daniel, B., Halbrook, R.S., Kloepper-Sams, P., Moore, M.N., Peakakk, D.B. (1992). The conceptual basis of the biomarker approach. In: Peakall DB and Shugart LR (Eds.) Strategy for biomarker research and Application in the Assessment of Environmental Health. Lewis, Boca Raton, Florida.

[32] Sarà, M., Vacelet, J. (1973). Ecologie des Démosponges. In Traté de Zoologie 3(1), Spongiaires, edited by P.-P. Grassé. Paris: Masson. pp. 462-576.

[33] Reiswig, H.M. (1971). Particle feeding in natural populations of three marine demosponges. *Biological Bulletin* 141: 568-591.

[34] Turon, X., Galera, J., Uriz, M.J. (1997). Clearance rates and aquiferous systems in two sponges with contrasting life-history strategies. *Journal Experimental Zoology* 278: 22-36.

[35] Ribes, M., Coma, R., Gili, J.M. (1999). Natural diet and grazing rate of the temperate sponge *Dysidea avara* (Demospongiae, Dendroceratida) throughout an annual cycle. *Marine Ecology Progress Series* 176: 179-190.

[36] Patel, B., Balani, M.C., Patel, S. (1985). Sponge "sentinel" of heavy metals. *Science of Total Environment* 41: 143-152.

[37] Olesen, T.M.E., Weeks, J.M. (1994). Accumulation of Cd by the marine sponge *Halichondria panicea* Pallas: Effects upon Filtration Rate and its relevanced for biomonitoring. *Bulletin Environmental Contamination Toxicology* 52: 722-728.

[38] Hansen, I.V., Weeks, J.M., Depledge, M.H. (1995). Accumulation of copper, zinc, cadmium and chromium by the marine sponge *Halichondria panicea* Pallas and the implications for biomonitoring. *Marine Pollution Bulletin* 31(1-3): 133-138.

[39] Cebrian, E., Martí, R., Uriz, M.J., Turon, X. (2003). Sublethal effects of contamination on the mediterranean sponge *Crambe crambe*: metal accumulation and biological responses. *Marine Pollution Bulletin* 46: 1273-1284.

[40] Cebrian, E., Martí, R., Agell, G., Uriz, MJ. (2006) Response of the Mediterranean sponge *Chondrosia reniformis* Nardo to heavy metal pollution. *Environmental Pollution* 141: 452-458.

[41] Turon, X., Tarjelo, I., Uriz, M.J. (1998). Growth dynamics and mortality of the encrusting sponge *Crambe crambe* (Poecilosclerida) in contrasting habitats: correlation with population structure and investment in defence. *Functional Ecology* 12: 631-639.

[42] Nickel, M., Brümmer, F. (2003). In vitro sponge fragment culture of *Chondrosia reniformis* (Nardo, 1847). *Journal of Biotechnology* 100(2): 23 147-159.

[43] Atkinson, B.G., Walden, D.B. (1985). Changes In Eukaryotic GeneExpression In Response To Environmental Stress. Academic Press, Orlando, FL, USA.

[44] Parsell, D.A., Lindquist, S. (1993). The function of heat shock proteins in stress tolerance: Degradation and reactivation of damaged proteins. *Annual Revue Genetics* 27: 437-496.

[45] Sanders, B.M. (1993). Stress proteins in aquatic organisms: An environmental perspective. *Critical Reviews in Toxicology* 23: 49-75.

[46] Sarkar, A., Ray, D., Shrivastava, A.N., Sarker, S. (2006). Molecular Biomarkers: Their significance and application in marine pollution monitoring. *Ecotoxicology* 15: 333-340.

[47] Brown, R.J., Galloway, T.S., Lowe, D., Browne, M.A., Dissanayake, A., Jons, M.B., Depledge, M.H. (2004). Differential sensitivity of three marine invertebrates to copper assessed using multiple biomarkers. *Aquatic Toxicology* 66: 267-278.

[48] Cajaraville, M.P., Bebianno, M.J., Blasco, J., Porte, C., Sarasquete, C., Viarengo, A. (2000). The use of biomarkers to assess the impact of pollution in coastal environments of the Iberian Peninsula: a practical approach. *The Science of the Total Environment* 247: 295-311.

[49] Sanders, B.M. (1990). Stress proteins: Potential as multitiered biomarkers. In McCarthy J, Shugart L, eds, Biological Markers Of Environmental Contamination. CRC Publishers, Boca Raton, FL, USA, 135-161 pp.

[50] Müller, W.E.G., Koziol, C., Kurelec, B., Dapper, J., Batel, R., Rinkevich, B. (1995). Combinatory effects of temperature stress and nonionic organic pollutants on stress protein (hsp70) gene expression in the freshwater sponge *Ephydatia fluviatilis*. *Environmental Toxicology Chemistry* 14: 1203-1208.

[51] Wiens, M., Koziol, C., Hassanein, H.M.A., Batel, R., Schöder, H.C., Müller, W.E.G. (1998). Induction of gene expression of the chaperones 14-3-3 and HSP70 by PCB118 (2,3', 4,4'5-pentachloro-biphenyl) in the marine sponge *Geodia cydonium*: Novel biomarkers for polychlorinated biphenyls. *Marine Ecology Progress Series* 165: 247-257.

[52] Müller, W.E.G., Batel, R., Lacorn , M., Steinhart, H., Simat, T., Lauenroth, S., Hassanein, H., Schöder, H.C. (1998). Accumulation of cadmium and zinc in the marine sponge *Suberites domuncula* and its potential consequences on single-strand breaks and on expression of heat-shock protein: a natural field study. *Marine Ecology Progress Series* 167:127-135.

[53] Agell, G., Uriz, M.J., Cebrian, E., Marti, R. (2001). Does stress proteins induction by copper modify natural toxicity in sponges? *Environmental Toxicology and Chem*istry 20 (11): 2588-2593

[54] Lindquist, S. (1986). The heat shock response. *Ann Rev Biochem* 55: 1151-1191.

[55] Lewis, S., May, S., Donkin, M.E., Depledge, M.H. (1998). The influence of copper and heat-shock on the physiology and cellular stress response of *Enteromorpha intestinalis. Marine Environmental Research* 46: 421-424.

[56] Nadeau, D., Corneau, S., Plante , I., Morrow, G., Tanguay, R.M. (2001). Evaluation for Hsp70 as a biomarker of effect of pollutants on the earthworm *Lumbricus terrestris. Cell Stress Chaperones* 6(2): 153–163.

[57] Köhler, H.R., Triebskorn, R., Stöcker, W., Kloetzel, P.M., Alberti G. (2004). The 70 kD heat shock protein (hsp 70) in soil invertebrates: A possible tool for monitoring environmental toxicants. *Journal Archives of Environmental Contamination and Toxicology* 22: 334-338

[58] Cochrane, B.J., Irby, R.B., Snell, T.W. (1991). Effects of copper and tributyltin on stress protein abundance in the rotifer *Brachionus plicatilis. Comparative Biochemistry and Physiology Part C* 98: 385-390.

[59] Cebrian, E., Uriz, M.J., Turon, X. (2007). Sponges as biomonitors of heavy metals in spatial and temporal surveys in northwestern Mediterranean: multispecies comparison. *Environmental Toxicology and Chemistry*

[60] Connor, P.M. (1972). Acute toxicity of havy metals to some marine larvae. *Marine Pollution Bulletin* 3: 190-192.

[61] Evenden, J., Depledge, M.H. (1997). Genetic susceptibility in Ecosystems: The Challenge for Ecotoxicology. *Environmental Health Perspectives* 105 (4): 1-12.

[62] Martin, M., Osborn, K.E., Billig, P., Glickstein, N. (1981). Toxicities of Ten Metals to *Crassostrea gigas* and *Mytilus edulis* embryos and cancer magister larvae. *Marine Pollution Bulletin* 12(9): 305-308.

[63] His, E., Geffard, O., De Montaudouinm, X. (1999). A comparison between oyster (*Crassostrea gigas*) and sea urchin (*Paracentrotus lividus*) larval bioassays for toxicological studies. *Water Research* 33(7): 1706-1718.

[64] Cebrian, E., Uriz, M.J. (2007) b. Contrasting effects of heavy metals and hydrocarbons on larval settlement and juvenile survival in sponges. *Aquatic Toxicology* 81: 137-143.

[65] Wu, R.S.S., Lam, R.K.S., Bingsberg, Z., 1997. "A settlement inhibition assay with cyprid larvae of the barnacle *Balanus amphitrite*." *Chemosphere* 35(9):1867-1874.

[66] Reichelt-Brushett, A.J., Harrison, P.L. (2000). The effect of copper on the settlement success of larvae from the scleractinian coral *Acropora tenuis. Marine Pollution Bulletin* 41(7-12): 385-391.

[67] Negri, A.P, Heyward, A.J. (2001). Inhibition of coral fertilization and larval metamorphosis by tributyltin and copper. *Marine Environmental Research* 51: 17- 27.

[68] Bellas, J., Vázquez, E., Beiras, R. (2001). Toxicity of Hg, Cu, and Cr on early developmental stages of *Ciona intestinalis* (Chordata, Ascidiacea) with potential application in marine water quality assessment. *Water Research* 55(12): 2905-2912.

[69] Blanquer, A., Uriz, M.J., Caujapé-Castells, J. 2007. Avoidance of inbreeding and clonality in a population of *Scopalina lophyropoda*, a sponge with short range larval dispersal and frequent fission events. *Marine Ecology Progress Series*, Submitted.

[70] Schindler, D.W. (1996). Ecosystems and Ecotoxicology: A personal perspective: In Ecotoxicology. A hierarchial Treatment, eds. MC Newman and CH Jagoe. Chapter 13 pp409-422. CRC Press, Boca Raton., Florida USA

[71] Honda, K., Nasu, T., Tatsukawa, R. (1986). Seasonal changes in mercury accumulation in the black-eared kite, *Milvus migrans lineatus. Environmetal Pollution* 42: 325-334.

[72] Hare, L. (1992). Aquatic insects and trace metals: bioavailability, bioaccumulation and toxicity. *Critical Reviews in Toxicology* 22: 327-369.

[73] Rainbow, P.S., Scott AG, Wiggins EA, Jackson RW (1980). Effect of chelating agents on the accumulation of Cadmium by the barnacle *Semibalanus balanoides* and complexation of soluble Cd, Zn and Cu. *Marine Ecology Progress Series* 2: 143-152.

[74] Perez, T., Longe, D., Schembri, T., Rebouillon, P., Vacelet, J. (2005). Effects of 12 years' operation of a sewage treatment plant on trace metal occurrence within a Mediterranean commercial sponge (Spongia officinalis, Demospongiae). *Marine Pollution Bulletin* 50: 301-309.

[75] Blackmore, G., 2001. Interspecific variation in heavy metal body concentrations in Hong Kong marine invertebrates. *Environmental Pollution* 114: 303–311

[76] Campanella, L., Conti, M.E., Cubadda, A., Sucapane, C. (2001). Trace metals in seagrass, algae and mollusc from an uncontaminated area in the Mediterranean. *Environmental Pollution* 111: 117-126.

[77] Agell, G., Turon, X., De Caralt, S., López, S., Uriz, MJ. (2004). Molecular and organismal biomarkers of copper pollution in the ascidian *Pseudodistoma crucigaster. Marine Pollution Bulletin* 48: 759-767.

[78] Rinkevich, B., Loya, Y. (1977). Harmful effects of chronic oil pollution on a Red Sea Scleractinian Coral population. Proc. 3rd Int. Coral Reef Symposium, Miami, Florida, pp. 585-591.

[79] Riisgård, H.U. (1979). Effect of copper on volume regulation in the marine flagellate *Dunaliella marina. Marine Biology* 50: 189-193.

[80] Correia, A.D., Costa, M.H., Ryan, K.P., Nott, J.A. (2002). Studies on biomarkers of copper exposure and toxicity in the marine amphipod *Gammarus locusta* (Crustacea): I. Copper-containing granules within the midgut gland. *Journal of the Marine Biological Association of the UK*, 82: 827-834.

[81] Perez, T., Longet, D., Schembri,.T., Rebouillon, P., Vacelet, J. (2005). Effects of 12 years' operation of a sewage treatment plant on trace metal occurrence within a Mediterranean commercial sponge (*Spongia officinalis*, Demospongiae). *Marine Pollution Bulletin* 50: 301-309.

[82] Berthet, B., Mouneyrac, C., Perez, T., Amiard-Triquet, C. (2005). Metallothionein concentration in sponges (*Spongia officinalis*) as a biomarker of metal contamination. *Comparative Biochemistry and Physiology Part C. Toxicology Pharmacology* 141(3): 306-13.

[83] Duquesne, S., Matthias, L., Birda. D.J. (2004). Sub-lethal effects of metal exposure: physiological and behavioural responses of the estuarine *bivalve Macoma balthica. Marine Environmetal Research* 58 (2-5): 245-250.

[84] Perez T, Vacelet J, Rebouillon P (2004). In situ comparative study of several Mediterranean sponges as potential biomonitors of heavy metals. In: Pansini, M., Pronzato, R. Bavestrello, G., Manconi, R. (Eds.) Sponge Science in the New Millenium. Bolletino dei Musei e degli Instituti Biologici dell'Universitá di Genova 68: 517-525.

[85] Evans, D.W., Dodoo, D.K., Hanson, P.J. (1993). Trace elements concentrations in fish livers. Implications of variations with fish size in pollution monitoring. *Marine Pollution Bulletin* 26 (6): 329-334.

[86] Catsiki, V.A., Strogyloudi, A. (1999). Survey of metal levels in common fish species from Greek waters. *The Science of the Total Environments* 237: 387-400.

[87] Cima, F., Ballarin, L., Bressa, G., Burighel, P. (1998). Cytoskeleton alterations by tributyltin (TBT) in tunicate phagocytes. *Ecotoxicology Environmental Safety* 40: 160-165.

[88] Ng, T.Y-T, Keough, M.J. (2003). Delayed effects of larval exposure to Cu in the bryozoan *Watersipora subtorquata. Marine Ecological Progress Series* 257: 77-85.

[89] Peckenick, J.A., Wrendt D.E., Jarrett, J.N. (1998) Matamorphosis is not a new beginning. *BioScience* 48: 901-910.

[90] Soazig, L., Marc, L. (2003). Potential use of the levels of the mRNA of a specific metallothionein isoform (MT-20) in mussel (mytilus edulis) as a biomarker of cadmium contamination. *Marine Pollution Bulletin* 46: 1450-1455.

[91] Roméo, M., Hoarau, P., Garello, G., Gnassia-Barelli, M., Girard, J.P. (2003). Mussel transplantation and biomarkers as useful tools for assessing water quality in the NW Mediterranean. *Environmental Pollution* 122: 269-378.

[92] Geffard, A., Amiard-Triquet, C., Amiard, J.C. (2005). Do seasonal changes affect metallothionein induction by metals in mussels *Mytilus edulis*? *Ecotoxicology and Environmental Safety* 61: 209-220.

[93] Etxeberria, M., Cajaraville, M.P., Marigomez, I. (1995). Changes in digestive Cell Lysosomal Structure in Mussels as Biomarkers of environmental stress in the Urdaibai Estuary (Biscay Coast, Iberia Peninsula). *Marine Pollution Bulletin* 30(9): 599-603.

[94] Riveros, A., Zúñiga, M., Hernández, A., Camaño, A. 2002. Cellular biomarkers in native and transplanted populations of the mussel Perumytilus purpuratus in the Intertidal zones of San Jorge Bay, Antofagasta, Chile. *Archives Environmental Contamination and Toxicology* 42: 303-312.

[95] Fagotti, A., Di Rosa, I., Simoncelli, F., Pipe, R.K., Panara, F., Pascolini, R. (1996). The effects of cooper on actin and fibronectin organization in *Mytilus galloprovincialis* Haemocytes. *Developmental and Comparative Immunology* 20(6): 383-391.

[96] Kontreczky, C., Farkas, A., NemcsoK, J., Salanki, J. (1997). Short- and long-term effects of deltamethrin on filtering activity of freshwater mussel (*Anodonta cygnea* L). *Ecotoxicology Environmental Safety* 38(3): 195-9.

[97] Gómez-Mendikute, A., Cajaraville, M.P. (2003). Comparative effects of cadmium, copper, paraquat and benzo[a]pyrene on the actin cytoskeleton and production of reactive oxygen species (ROS) in mussels haemocytes. *Toxicology in vitro* 17: 539-546.

[98] Brown, M.T., Depledge, M.H. (1998). Determinants of trace metal concentration in marine organisms. Metal metabolisms in Aquatic Environmnets. Edited by W.J Langston MJ. Bebianno. Chapman ans Hall London.

In: Aquatic Toxicology Research Focus
Editor: Elias P. Svensson, pp. 119-144
ISBN: 978-1-60456-192-0

Chapter 6

FINE SCALE SPATIAL AND TEMPORAL VARIATION IN THE STRUCTURE AND FUNCTION OF AQUATIC MICROBIAL COMMUNITIES

Kurt A. Smart [1]***, Hsuanhua L. Smart*** [2] ***and Colin R. Jackson*** [1]

1.The University of Mississippi, Department of Biology,
Shoemaker Hall, University, MS 38677, USA
2The University of Mississippi, Department of Mathematics,
Hume Hall, University, MS 38677, USA

ABSTRACT

Aquatic ecosystems have been heavily impacted by manufactured chemicals and other anthropogenic materials with dramatic effects on the microbial communities of these systems. However, few studies have examined how fine scale changes in environmental conditions resulting from natural or man-made forces can alter the structure, diversity, and function of aquatic microorganisms. We performed a number of studies that examined the influence of fine scale factors such as particle size or spatial location on microbial communities in freshwater systems. Bacterial communities associated with different sizes of particles in wetland sediment were characterized using 16S rRNA gene sequencing, which revealed fundamental differences in the structure of communities associated with very fine (0.063-0.25 mm diameter) and fine (0.25-1 mm) particles. Similarly, microbial activity was also influenced by particle size with different activities of lignin and cellulose degrading enzymes associated with different sized particles. Fine (cm) scale spatial variation in microbial activity was also seen on decomposing leaves in a stream, and spatial patterns in cellulose degrading enzymes suggest that the process of organic matter decomposition shows high spatial heterogeneity. These findings demonstrate that aquatic microbial communities vary both structurally and functionally over finer scales than are usually considered and these fine-scale interactions should be considered when assessing microbial communities for effects of or treatment and removal of environmental stresses.

INTRODUCTION

Aquatic ecosystems have been significantly impacted by anthropogenic pollutants such as heavy metals, agricultural pesticides, and industrial solvents, to name but a few, and there is increasing interest in removing, or at the very least dampening the effects of such activities and restoring more natural processes. Microorganisms have received a great deal of attention in these matters. Interest in microbial community structure and function in aquatic ecosystems as a whole has been increasing dramatically, but especially regarding the impact of man-made pollutants as microorganisms are critical to the understanding and removal of such chemicals from these environments. The significance of microorganisms in biogeochemical cycles and ecosystem function has been noted since Lindeman [1] placed the bacterial 'ooze' at the center of his trophic cycle, and more recently microorganisms have been used successfully to remove or transform toxic or potentially hazardous substances through bioremediation [2]. But there is still much to be learned about microbial community structure and function in aquatic environments, both impacted and pristine, especially at fine-scales which have historically been difficult to study. Recent developments in molecular and biochemical techniques are helping to overcome those difficulties and are beginning to shed some light on microbial processes at very fine scales [3]. These new methods and techniques are allowing more accurate, detailed descriptions of microbial communities and their activities, and are revealing how diverse these communities are both in structure and function.

Microorganisms respond to localized, fine-scale changes in their environment, both spatially and temporally, and it is important that these responses and interactions be understood more thoroughly [4]. Knowledge of such fine-scale processes are critical in understanding larger scale phenomenon, such as ecosystem or global level cycles, as fine-scale interactions and activities of microorganisms combine together to form these larger scale processes. From a hierarchal perspective it is important to also consider multiple levels of organization (where organizational levels are created from lower levels but also act to create higher levels of organization) as the relationships between levels are what provide the insights into the function of the system [5]. Lower levels of hierarchal organization (i.e. fine scales) are highly variable and operate at high rates, but those rates can be averaged out when looking at the system from higher levels [6]. For example, Krembs *et al.* [7], looking at patchiness in planktonic organisms, argue that concentration dependent rate processes may be underestimated by failing to take into account fine-scale spatial variability (i.e. by overlooking processes and patterns at fine scales, any variability that may be present is missed, making it impossible to know what is actually occurring to create the larger scale system and its function). Until recently, the ability to observe and measure fine-scale microbial processes was very limited.

Bacteria have often been regarded as a single entity in aquatic environments due to the difficulties involved in studying them in any detail. Early studies of environmental microbiology relied upon culture-based methods for describing microbial communities, though it is now commonly accepted that these methods greatly underestimate both the microbial diversity and viable counts of microbial populations. The vast majority of prokaryotic species could not be cultured and most still have yet to be cultured using current culturing techniques. Culture-based methods often select for organisms which either do not make up a significant portion of the community or do not contribute a great deal to the activity of the community, and provide a skewed representation of what is actually present. Rapid advances in molecular and biochemical methods have helped to overcome the bias involved by culture dependent methods and have allowed microbial communities to be described more accurately [3]. Molecular techniques have also become increasingly sensitive, enhancing the ability for measurement on finer and finer scales.

The application of thc polymcrasc chain reaction (PCR) has been the most significant development for dealing with environmental microorganisms, allowing replication of minute amounts of environmental DNA into quantities large enough to be analyzed. The use of fluorescently labeled compounds has also greatly benefited environmental microbiology, detecting organisms and measuring a wide range of enzymatic or metabolic processes at very high sensitivities. Fluorescent *in-situ* hybridization (FISH) combined with confocal laser microscopy has enabled direct observations of microbial community structure, providing valuable data on the distributions of microbes [8]. Along with molecular tools and advances in microscopy, the development of new biochemical assays and instruments have proved helpful in measuring fine-scale activities and function [9]. Microsensors have also been widely implemented in determining microenvironmental conditions and processes, measuring chemical concentrations and gradients over very fine scales [10]. The use of such methods and several more are discussed in this chapter relating to the study of fine-scale heterogeneity in microbial communities.

Microorganisms are not found to be distributed homogeneously in any aquatic environment. Microbial species composition and abundance in lakes are known to vary with depth depending upon physical (e.g. light availability, temperature), chemical (e.g. O_2, CO_2), and biological (e.g. symbiotic or predator-prey relationships) characteristics [11]. They are also known to vary between littoral and pelagic habitats, and seasonally [11]. In lotic systems the distribution varies more with horizontal rather than vertical profiles due to differences in hydrological conditions [12]. However, our knowledge of that variability is based upon observations at spatial scales of meters to tens of meters and temporal scales of weeks to months. Microorganisms have more recently been shown to be heterogeneously distributed throughout the water column, on submerged surfaces, and in aquatic sediments over much finer spatial and temporal scales. The following discussion focuses on the effects of fine-scale environmental variation on aquatic microbial communities and how it may have significant implications for understanding the role of manufactured chemicals and other anthropogenic and natural materials in aquatic environments. We follow the discussion by presenting two examples of studies we performed examining fine-scale patterns in the structure and function of microbial communities in aquatic environments.

Fine Scale Patterns in the Water Column of Aquatic Environments

Fine-scale patterns and processes in the water column of aquatic environments are beginning to shed some light on the function of these ecosystems, and it is likely these patterns will provide important insights into understanding the transformation and cycling of anthropogenic pollutants. Microorganisms are found to be very heterogeneously distributed in the water column of aquatic environments at scales much smaller than previously thought.

Bacterioplankton abundance has been shown to vary on a scale of several millimeters [13], and at even finer scales bacteria have been shown to form aggregates or associations on or near suspended organic particles [14] and phytoplankton [15], or with other microorganisms [16]. Microbial communities have even been observed to be patchily distributed on the surfaces of particulate matter, demonstrating how small the microenvironments that they inhabit are [17].

There appear to be several reasons for this non-uniform distribution and behavior at such fine scales. The most apparent reason is the availability of nutrients. Phytoplankton, for example, are not distributed evenly throughout the water column and have recently been shown to be heterogeneously distributed in these environments, forming thin layers or concentrated sheets that may be only centimeters thick [18]. These sheets of phytoplankton release photosynthetic byproducts that attract high concentrations of bacteria, forming dense, dynamic communities surrounded by relatively uninhabited waters [15]. Particulate organic matter (POM) also contributes to fine-scale spatial heterogeneity in the water column of aquatic systems as POM is a prime source of nutrients and energy for aquatic microorganisms. POM is the remains of larger organic matter which has been partially degraded by bacteria and fungi, and in comparison to the surrounding water column is rich in nutrients. It can also be highly concentrated with metals, accounting for up to 79% of the metal in aquatic systems with concentrations up to 35,000 fold higher associated with particulate matter than in the water column [19]. Dense accumulations of highly active microorganisms gather on and around POM to take advantage of the rich substratum, greatly increasing the overall spatial heterogeneity of microbial biomass and activity at fine scales [20]. Over fine scales the distribution of detrital particles themselves are not uniform, varying greatly in size, shape, and nutrient composition [20]. All of these particles are a much more plentiful source of nutrients and carbon for bacteria and fungi than scavenging in the surrounding water and leads to the formation of dense areas of active and diverse microbial communities.

Another possible advantage to the patchy distribution of microorganisms may be defense against predation. Predation by protists can significantly control bacterial populations, and grazing pressure from protists has been shown to alter the dominant cell morphology within bacterial communities [21]. Some microbial species may even form large colonies to avoid or reduce the impact of predation [16]. Residing on large particles may provide similar benefits, although other studies have found that protists are also closely associated with aggregates and that grazing pressure on aggregates is quite high [14]. Higher bacterial growth rates on POM may be able to overcome the elevated grazing pressures or may be a response to them. It is also possible that particle-associated bacterial communities may form biofilms, producing extracellular polymeric substances (EPS) which may offer some grazing protection. More attention is being given to understanding such interactions, but it is highly speculative and too early to draw any general conclusions.

The majority of particulate matter-microbial community research has focused on particulate matter towards the larger end of the size scale. These large particles, or macroaggregates, dubbed ‘marine snow’, ‘lake snow’, and more recently ‘river snow’, were the first to be studied in detail as they are easily visible with the naked eye (typically ≥0.5 mm in diameter). However, macroaggregates are formed from finer particles which are bound together with transparent exopolymer particles produced by algae, diatoms, and bacteria [22]. These aggregates are composed of a wide variety of smaller particles, such as diatom frustules, dead phytoplankton cells, fecal matter, decomposing plant matter, or heavy metals and manufactured chemicals in impacted environments, and are highly variable in both nutrient content and availability. Nutrients in POM and other particles are not homogenously distributed, and not all particles are equal in nutritional quality or availability; fine-scale heterogeneity of attached microorganisms is therefore likely to be increased by the surfaces of the particles they are attached to.

A variety of forces, both physical and biological, act to create very steep physical and chemical gradients associated with POM as well as with other aggregates. Respiration of heterotrophic bacteria colonizing both the interior and surface of particles can create steep oxygen gradients [23] and the diffusion of gases and dissolved nutrients is often constrained within the inner core of aggregates [24]. pH gradients resulting from heavy metal contamination may also lead to significantly different microenvironments within an aggregate [19]. Smith *et al.* [25] measured the activity of a wide range of hydrolytic enzymes on individual aggregates of marine snow collected off the California coast and found high variability in all of the enzymes tested, suggesting their are major differences among particles. Some of the variability in activity and chemical and physical gradients may be explained by the highly variable surface area of aggregates, which are more variable than commonly thought. Over very fine scales the surface of detritus and other aggregates can be very fractal in nature, which would mean that the microbial communities are not spread evenly across a perfect sphere as is often assumed [26]. Interstitial spaces may also be highly variable between particles [27] and these gradients may lead to differences in community structure as they create a wide range of microenvironments with dramatically different physical conditions.

Not all of the microenvironments associated with aggregates occur on or in the particle itself, however, such as the formation of nutrient plumes released from sinking POM. On the surface of POM, heterotrophic bacteria and fungi attached to the surface produce extracellular enzymes, hydrolyzing the POM and releasing nutrients which are then free to be taken up by the colonizing microbes. Some of the released nutrients may leak into the surrounding water column as not all of the nutrients released by this enzymatic degradation are always taken up by the colonizing microorganisms, and this released matter may attract planktonic organisms from the surrounding area [14]. As a particle sinks the fluid flow past the particle creates a trail, or plume, of released nutrients in the wake of the particle. These plumes have been observed to be several meters in length and remain intact for several minutes [28]. Plumes of this nature attract chemotactic bacteria, zooplankton, and protists which are thought to be well-adapted to life unattached to particles, and may play a role in attracting new inhabitants to particles [28].

All of the particles examined to date have been found to support highly abundant and diverse microbial communities, with the species composition of suspended particles being shown to be significantly different from that of free-living microbial populations. In a study of the Columbia River estuary, Crump *et al.* [29] compared the 16S rRNA gene sequences of aggregate-associated bacteria with those of free-living bacteria and observed major differences in freshwater, saltwater, and brackish environments. DeLong *et al.* [30] compared the 16S rRNA gene sequences of aggregate-associated bacteria with those of free-living bacteria in marine waters and found they were also very distinct: aggregate-associated species were typically members of the Cytophagales, Planctomycetes, and gamma-Proteobacteria, while the free-living species were all alpha-Proteobacteria. In a similar analysis in Lake

Constance alpha-Proteobacteria were found to be the dominant members of the macroaggregate-associated community along with the beta-Proteobacteria, although alpha-Proteobacteria were not detected on finer microaggregates [31]. This suggests that bacterial communities may be adapted to the microenvironments of particulate matter and possibly to specific niches within those microenvironments. 'River snow' also appears to have different aggregate-associated microbial communities [17] and the structure of microbial communities appears to differ between these environments, but the reasons for this are still unknown. From these observations it is likely that aggregates composed from anthropogenic sources will have their own unique microbial communities associated with them. We describe differences in bacterial community structure associated with different sizes of particles later in this chapter.

As well as supporting diverse communities of microorganisms, detrital particles are also described as 'hot-spots' of microbial activity, supporting microbial populations several orders of magnitude greater than the surrounding environment, with larger, more active individual cells [32]. Growth rates are much greater for attached bacteria due to the greater availability of nutrients and their elevated activity in response to these conditions [33]. Upon attaching to surfaces, planktonic bacteria up-regulate their activity and increase production of extracellular enzymes; enzymatic activity of attached bacteria can be orders of magnitude greater than planktonic bacteria [34]. Up-regulation occurs very suddenly, within minutes, and leads to rapid decomposition of the detritus and remineralization of nutrients [28]. There is a very dynamic structure to the fine-scale formation and degradation of macroaggregates as macroaggregates are degraded by highly active bacteria, releasing smaller particles; the smaller particles often then reassociate to reform larger particles. The degradation of aggregates effect the rate at which they sink, with larger particles sinking at a greater rate [35], and as particles are degraded and reformed rather rapidly it may reduce the overall sinking rate and allow more time for microbial remineralization of nutrients. This dynamic fine scale process has major implications for carbon cycling and sedimentation in aquatic environments and has significance at even larger scales.

Microorganisms on or near the surface of aggregates are closely associated and it is likely that competition for resources is high. There is some evidence that suggests microbial communities are functioning as consortia with interspecific interactions that enhance the ability of other microorganisms to colonize the substrate and take up nutrients at a greater rate, as Grossart *et al.* [36] found that the presence of certain bacterial species on aggregates enhanced the ability of other species to colonize the substrate while reducing the ability of others. It is also likely there are fine spatial and temporal scale successional patterns occurring in these microenvironments [37]. Aggregates are also likely 'hot-spots' for viral infection, as viruses are also closely associated with particulate matter, though the degree to which they are associated varies by both particle characteristics and environmental conditions [38]. Very little is known about competition, succession, mutualistic relationships, or bacterial-viral interactions on aggregates, but it seems likely these interactions play a significant role in the behavior of these communities considering the interactions seen in other closely associated microbial communities, such as biofilms.

Ever finer particles are being discovered and are revealing the behavior and effects of man-made and natural chemicals at very fine scales. Nanoparticles, particles measuring 10^{-9}-10^{-7} m in length and consisting of either natural or anthropogenic chemicals, have recently begun to receive attention due to their effects on water quality in aquatic systems [39]. Nanoparticles are formed from a variety of sources and it is likely they will reveal even more diverse and finer scale patterns in microbial community structure and function. More sensitive tools and methods are still being developed to examine these environments.

FINE SCALE PATTERNS IN MICROBIAL BIOFILMS

Microorganisms attached to surfaces in aquatic environments are most often present in much higher numbers than microorganisms unattached to surfaces [40], colonizing all available submerged surfaces and forming densely packed, complex, interactive communities (biofilms) composed of living cells and both organic and inorganic matter heterogeneously distributed in both horizontal and vertical directions and encased in EPS. Biofilms provide an advantage for microorganisms in stressful environments, such as those exposed to extreme temperatures or pH, and their function resembles that of multicellular organisms [41]. Very fine-scale interactions and dynamics, both temporal and spatial, occur within biofilms, between biofilms and the colonized substratum, and between biofilms and the surrounding environment, which exhibit a wide array of functions and metabolic processes and which play a significant role in nutrient and energy cycling in aquatic ecosystems [42]. Recently the fine-scale function of biofilms in regards to anthropogenic stress, particularly those of heavy metal pollution and toxicity, have begun to receive a great deal of attention as biofilms can accumulate high concentrations of these pollutants [43].

Biofilm formation is initiated when planktonic bacteria come into contact with and attach to surfaces, forming microcolonies and spreading both across the surface and upwards above the surface in a three-dimensional manner. This colonization and growth is spatially non-uniform as microcolonies approach each other forming diverse microenvironments in both the horizontal and vertical planes. Across the substratum heterogeneity in biofilm structure is caused by both the presence of other inhabitants and by the surface to which the biofilm is attached. Surfaces vary greatly in chemical and physical characteristics, and microbial communities respond to those differences. Jones *et al.* [44] found that the microbial communities on different artificial substrata in an estuary were similar in species composition during the early stages of biofilm formation but were different after a week of incubation. Their results suggest that attachment is at first random, but that characteristics of the substratum favor certain microbial inhabitants over others. At very fine scales surfaces can be highly heterogeneous and fractal in nature, similar to that of particulate matter, which affects not only the surface area available for colonization but also environmental conditions, such as flow, which can affect pH, nutrient availability, waste removal, and oxygen availability, among others [45, 46]. Fine-scale surface characteristics such as these can lead to large-scale impacts, including significant alterations in microbial productivity and respiration in stream ecosystems [47]. Anthropogenic factors are also likely to have large scale effects as the community composition of biofilms has been found to change in response to these stresses [48].

Close association among biofilm inhabitants and diverse microenvironments present within biofilms may be responsible for a number of complex and significant chemical transformations. The potential for Fe redox cycling was found to be possible in microenvironments within a biofilm along the interface between aerobic and anaerobic conditions [49]. This reaction was previously considered unlikely to occur in any natural environment. Okabe *et al.* [50] found that sulfate-reducing bacteria were dominant at the oxic-anoxic chemocline in activated sludge biofilms and formed a thin, highly active sulfate-reducing layer in the film. These microenvironments within biofilms enable sulfate-reducing bacteria to be active in an otherwise aerobic environment [51] and other metabolic processes can show similar patterns in biofilms [52, 53]. Analysis of acid-mine drainage microbial communities in Spain also found a diverse array of metabolic activities, with significant interactions both among subsurface biofilm inhabitants and between the biofilm surface and water column [54]. The macroenvironmental conditions in this system suggested there should have been high rates of iron oxidation, though that was not found to be the case.

The vertical dimension of biofilms creates diverse microenvironments and supports a wide range of metabolic activities that enable the processes described above. Biofilms are often several hundred micrometers thick and diffusion of gases and nutrients through the biofilm matrix are often restricted [55]. Microsensor measurements show that anoxic conditions can occur within 30 micrometers of the surface even if the surface is saturated with oxygen [56]. The different physical and chemical characteristics observed throughout the vertical dimension of biofilms are commonly found to allow tight coupling of reactions and cycles similar to that of the Fe redox reactions and sulfur cycles already mentioned, with great significance at larger scales. In addition to distinct functional processes at different levels within biofilms there are also often specific community compositions inhabiting these levels [55]. Even biofilms containing single-species and grown under laboratory conditions show differentiation of cells and a non-uniform distribution of function [57].

Other forces also act to create heterogeneity in both microbial community structure and activity within biofilms. Communication between microbial populations is a dominant factor, responsible for coordinating the activity of the tightly connected communities, as biofilms contain structures not seen in other microbial communities and are very detailed at fine scales [58]. Channels are common structures formed throughout biofilms, allowing water to flow through the film, transporting nutrients and wastes. The flow is often less than that at the surface of the biofilm and has been observed to flow both with and against the surface current [59], demonstrating the differences in physical forces acting upon organisms within biofilms compared to the surface. Quorum sensing chemicals are an important factor in determining biofilm structure and function and are more concentrated than chemical signals outside of biofilms, likely due to the decreased flow rates in biofilms and the high density of extracellular polymeric substances that encapsulate them, reducing the diffusion rate of the signal and localizing its effects [60]. The hydrophobic properties of many of the signal molecules may also allow the signals to remain localized [58]. Other gradients within the biofilm, such as pH gradients, also restrict the dispersion of signals by affecting the degradation rate of the chemical [58]. Restricted diffusion of signals in biofilms is another example of an extremely fine-scale process that could have a tremendous impact on larger scale functions.

A large portion of biofilms often consist of inactive or dormant cells that are not contributing to the activity measured at larger scales. Francoeur *et al.* [61] demonstrated this heterogeneity using fluorescently-labeled substrate to detect extracellular protease activity, observing a few highly active cells surrounded by a void of inactive, or 'dead', cells. Analysis of a biofilm in drinking water that received an addition of several substrates demonstrated similar spatial heterogeneity in microbial activity [62]. Heavy metal accumulation and transformation also shows high spatial heterogeneity within biofilms. EPS, other microbial polysaccharides, and cell surfaces attract metal ions from the environment, and biofilms may accumulate relatively high concentrations of them [43]. Some cells may accumulate metals for metabolic purposes whereas others may accumulate them without any desired function [63]. This accumulation and transformation may lead to significant differences in metabolic activity across very short distances.

Protozoan grazing of biofilms also affects the structure and function of biofilm bacterial communities and plays a major role in influencing fine-scale interactions within biofilms. Biofilms might be thought to offer protection against predation by protists, but protozoan grazing can dramatically affect biofilm population dynamics [64]. Grazing has been shown to increase heterogeneity of biofilms as protists preferentially graze on small patches in the biofilm, significantly reducing the local microbial populations [65]. Protists also appear to target areas that others have already grazed, emphasizing local grazing patterns [65]. The cause of this selective grazing is not yet known, but it is likely to have an effect on larger-scale processes.

Temporal scales are also important to consider as microorganisms present in younger, newly developing biofilms appear to be much more active than those in mature biofilms [62]. Conditioning of the substratum and the subsequent attachment of microbes occur at very fine temporal scales and growth and spread of colonies over the surface is rapid, with surfaces introduced to aquatic environments being densely covered in a few days. Over longer time intervals biofilm communities undergo successional changes as the biofilm develops and matures [66]. These changes in community composition are a result of fine-scale interactions and changes in microenvironmental conditions which enable a more diverse set of metabolic activities [67]. Microbial inhabitants of biofilms are closely linked to each other, often forming a consortium of interacting organisms, benefiting one another by co-metabolizing substrates or metabolizing each others waste products. They are very intricately detailed communities which appear to function almost as multicellular organisms and have become widely studied for their uses in bioremdiation of impacted ecosystems.

Fine Scale Variation Patterns in Sediments

Aquatic sediments share many of the characteristics of biofilms and suspended POM, and fine-scale heterogeneity of microbial community structure and function in sediments are similar. Sediments consist of a wide range of particles that are diverse in both physical and chemical characteristics, varying greatly in size, shape, density, organic and inorganic nutrient composition, and surface features [20]. The environment in which the sediment particles are located are also highly diverse and significantly impact the structure and activity of the microbial biota inhabiting them; river and stream sediments are markedly different than their lake and ocean counterparts, and tidal sediments have their own distinct properties due to the very different physical forces acting upon them [68 - 70]. River and stream sediments are constantly changing as sediments are pushed downstream and resuspended from the benthos; lake sediments fluctuate because of suspended particles settling out; and tidal sediments fluctuate in response to tidal currents at fine temporal scales.

Sediments are similar to biofilms in regard to their vertical profiles, where different processes occur at different depths and where microenvironmental variations support a diverse range of activities. Fine-scale measurements of contaminated marine sediments discovered that heavy metals were concentrated in bands ranging from 1-5 mm thick below the sediment surface, and two of the metals measured (Cu and Cd) fluctuated between the sediment and being remobilized into the water column, whereas the others remained concentrated in the sediment [71]. These interactions are likely to influence the microbial communities on or near the surface and their activity in cycling of these products. The inability of nutrients and gases to diffuse evenly through sediments also creates layers which are significantly different in composition of nutrients and energy [72]. In lotic systems, turbulent flow that disrupts sediments can greatly increase bacterial density and activity on fine-scales, especially at the sediment surface [73]. Fine-scale fluctuations in pore size and sediment topography also lead to highly diverse patterns in biogeochemical cycles at fine temporal (seconds) and spatial (cm) scales, increasing or decreasing the flow of water through sediments and altering the physical and chemical microenvironment [74].

Dynamics on the surface of particles in sediments are similar to the dynamics on suspended particles. Differences in activity have been measured between different size fractions of particles in sediments in a wide range of aquatic habitats, with finer particles in wetland and river sediments having greater bacterial biomass and productivity compared with coarser particles [75]. Particle size also influences extracellular enzyme activity and bacterial community structure, as we describe later in this chapter. Contaminants also may alter activity as heavy metals have been associated with decreased activity in aquatic sediments. Metals are also found to be preferentially associated with different sizes of particles, with finer sediments accumulating around 3x the concentration of metals than coarser, sandier particles [76].

Uneven distribution of nutrients and energy also results in fine scale heterogeneity in sediments. Particulate matter is not distributed homogeneously in sediments and is not distributed evenly over time. In aquatic systems organic matter is distributed in patches, as in the case of allocthonous plant matter [77] or decaying fish and plankton carcasses [78]. Many anthropogenic pollutants also accumulate in these tissues and likely contribute to their heterogeneous distribution [79]. Uneven distribution of organic nutrients in wetland sediments has been shown to be correlated with spatial heterogeneity in methane production [80], suggesting spatial patterns in methanogen communities. Fine-scale spatial heterogeneity may also arise due to contamination from agricultural or industrial run-off, as pollutants such as lead and mercury are often patchily distributed [81]. A study of heavy metal contaminated soils revealed that the composition and activity of microbial communities varied dramatically over fine scales, with the potential metabolic activity varying up to 10,000 fold over a single centimeter [82].

Invertebrate burrowing can also increase the heterogeneity of microenvironments in aquatic sediments, creating oxic conditions in layers of sediment that would otherwise be anaerobic [83]. Invertebrates, fish, and other macrofauna are also responsible for disturbing sediments and altering microenvironments, altering fine-scale distributions of nutrients and microorganisms while feeding or traveling along the benthos [84]. These burrows and disturbed sediments have greatly elevated O_2, OM, and enzyme activities associated with them compared with nearby undisturbed sediments [85]. Variability at fine scales may also arise due to aquatic plants and their root structure, with bacterial populations in the rhizosphere differing from those in the surrounding sediment [86]. Fine-scale interactions are common within the rhizosphere as a variety of factors influence nutrient availability and environmental conditions over small areas [87].

THE EFFECTS OF PARTICLE SIZE ON BACTERIAL COMMUNITY STRUCTURE AND ACTIVITY IN WETLAND SEDIMENTS

As stated previously, particle size influences bacterial biomass and productivity in aquatic sediments [75]. At relatively coarse scales, sediment particle size also influences the structure of the associated microbial community. Particles in aquatic environments can be designated as either coarse particulate organic matter (CPOM, >1 mm diameter) or fine particulate organic matter (FPOM, <1 mm diameter) based upon size [88], and there are differences in the microorganisms associated with these classes of particles. Fungi are thought to dominate the decomposition of CPOM, whereas the smaller size of FPOM limits the growth of fungal hyphae and bacteria tend to be more abundant [89]. However, fewer studies have examined how bacterial community structure changes with decreasing size of FPOM, i.e. how fine scale changes in particle size might influence the attached bacterial community. While FPOM refers to all particles below 1 mm in diameter, it essentially consists of particles that are the remains of decomposing CPOM, and smaller (generally <100-200 μm diameter) particles that may come from a variety of sources, such as the precipitation and flocculation of dissolved material [90]. Given these differences, it seems reasonable to hypothesize that there may be differences in the bacterial communities associated with different sizes of FPOM, and basic DNA-DNA dot blot hybridizations have suggested that this is indeed the case [91]. The DNA-DNA hybridization technique does not directly assess community structure or diversity, however, and interpretation of patterns can be problematic [92]. Thus, we used 16S rRNA sequencing techniques to investigate the structure and diversity of the bacterial community present on different sizes of particles in wetland sediments. A companion study examined microbial extracellular enzyme activity associated with the different sizes of particles.

Sediment samples were collected from a cypress-tupelo swamp in southeastern Louisiana, USA, and sieved into two size ranges of particles (38-250 and 250-1000 μm diameter). DNA was extracted from each size range of particles using a modified version of a protocol originally designed for soil [93] but also used for geothermal spring sediment [94]. Residual humic contamination was subsequently removed from extractions using Sepharose 4B spin columns [95]. DNA extracted from both sizes of particles was used as the template in PCR amplifications [94] using primers specific for Domain Bacteria. These primers (Bac8f and Univ1492r) amplify positions 8-1492 of the 16S rRNA gene in Bacteria [96]. PCR products were purified, cloned into artificial plasmid vectors, and clone libraries (one for each size of particles) generated following procedures described previously [97]. The inserts from the first 96 clones obtained in each library were amplified and digested with three separate restriction enzymes (EcoRI, RsaI, HaeIII; 2 h digest at 37°C). After each digest the resulting restriction patterns were visualized on agarose gels. Clones in the same library that showed the same restriction patterns for all three restriction enzymes were assumed to have similar 16S rRNA genes inserted and were grouped together as a single phylotype [94, 97].

The frequency of each phylotype in each clone library was used to estimate the total bacterial richness present on particles of each size range using the non-parametric abundance-based estimator (S_{Chao1}) of Chao [98]. Calculations were performed using a web interface [99]. For each dominant phylotype, a portion of the 16S rRNA gene insert in one representative clone was sequenced, using the original Bac8f primer. This sequencing protocol yielded approximately 600 bp of sequence information, corresponding to roughly the 50-650 bp region of the 16S rRNA gene. The partial 16S rRNA sequences obtained were compared to those in GenBank to identify closest relatives and subsequently imported into the ARB software package [100]. Sequences were aligned automatically (FastAligner) and alignments checked manually. Because the partial 16S rRNA gene sequences obtained in this study represented a number of diverse lineages, novel phylogenetic trees were not constructed as this might produce erroneous results compared to existing relationships for major bacterial taxa. Rather, ARB's "Quick add by parsimony" function was used to incorporate the sequences into existing phylogenetic trees in the ARB database. 16S rRNA sequences in the existing trees that were unrelated to those obtained in this study were subsequently removed without altering tree topology, so that phylogenetic trees showing only the lineages specific to this study were obtained. This treeing procedure has been used previously to minimize distortions to established tree topologies that might arise from a phylogenetic analysis of short, highly divergent sequences [97].

When S_{Chao1} was calculated by repeated random sampling from each library, estimates of the total bacterial phylotype richness in the original community were 177 phylotypes associated with larger particles and 288 phylotypes associated with the finer particles. In each case, S_{Chao1} began to approach a stable asymptotic value although this was not pronounced for the clone library derived from the finer sediment particles (Figure 1). This suggests that while the estimated richness for the community associated with primary particles is a reasonable approximation, the value of S_{Chao1} determined for the community associated with the finer, secondary particles is probably an underestimate. Thus, not only does the structure of the bacterial community differ between sediment particles of different size ranges, but overall community properties such as species diversity appear to differ also, with finer particles harboring substantially more diverse bacterial communities. This supports previous observations of increased bacterial community heterogeneity with decreasing FPOM particle size in aquatic sediments, and may be the result of increasing surface area/volume ratios on finer particles, which would result in greater relative habitat area and more available ecological niches [91]. The greater chemical complexity of secondary particles could also account for increased bacterial diversity on those particles.

Most 16S rRNA sequences showed low (<97 %) similarity to those already in the GenBank/EMBL/DDBJ databases. The majority of phylotypes occurred just once in each library, and repeated phylotypes were relatively rare. Across both sizes of particles, the sequenced phylotypes represented thirteen major bacterial lineages as well as a number of other taxa that have only been described from molecular studies. Sequences affiliated with the delta- and beta-Proteobacteria were the most abundant sequences found in the clone libraries for both sizes of particles, although the delta-Proteobacteria were more dominant in the library generated from 250-1000 μm particles (Figure 2). Both of these groups, along with the alpha-Proteobacteria are typically dominant in freshwater sediment communities [101]. Sequences affiliated with the alpha-Proteobacteria were abundant in the clone library derived from the larger particles but were comparatively rare in the finer particle library (Figure 2).

Similar differences between the two clone libraries were also seen for the Chlorobi (green sulfur bacteria), Nitrospira, and Spirochaetes, the latter being undetected in the clone library derived from 38-250 μm particles. In contrast, the library derived from these finer particles contained more 16S rRNA sequences affiliated with the Verrucomicrobia, Chloroflexi (green non-sulfur bacteria), and various uncultured bacterial lineages (Figure 2). Thus, the 16S rRNA sequences derived from the different size ranges of particles showed appreciable differences in their phylogenetic affiliation at broad taxonomic levels.

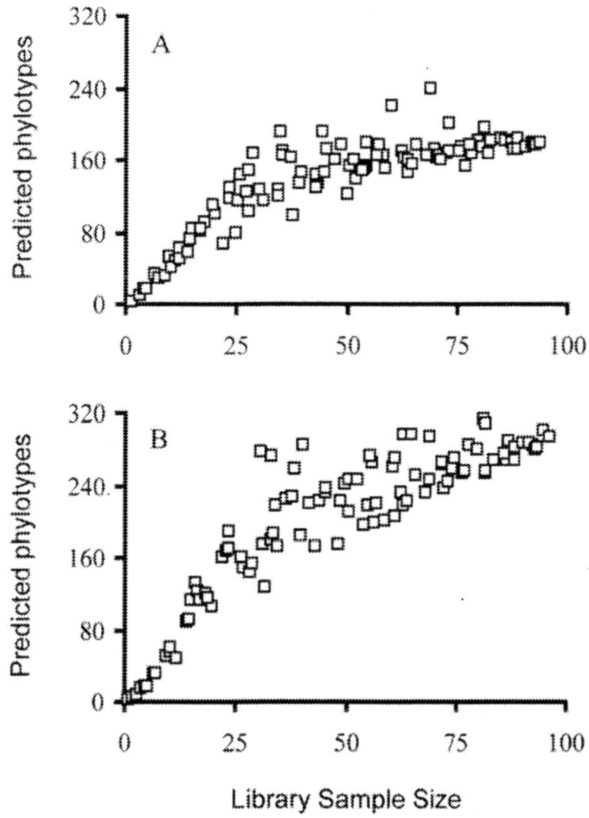

Figure 1. Predicted species (phylotype) richness as determined from analyses of 16S rRNA clone libraries generated from bacterial communities associated with two different sizes of particles (A: 250-1000 μm diameter; B: 38-250 μm diameter) in wetland sediments. Finer particles (B) harbor substantially more diverse bacterial communities than coarser particles (A).

The identity of the dominant phylotypes in each clone library also differed substantially. 16S rRNA sequences affiliated with members of the Bacteroides and the Methylococcales (gamma-Proteobacteria), a group of methanotrophic Bacteria, were the most abundant individual phylotypes in the clone library derived from larger particles, and these phylotypes each accounted for 5% of the clones sequenced. Other specific phylotypes that were abundant in this clone library were affiliated with the Rhodocyclales (beta-Proteobacteria), Pseudomonadales (gamma-Proteobacteria), and Myxococcales (delta-Proteobacteria). None of these orders of Bacteria were represented by abundant phylotypes in the clone library derived from finer particles. As might be expected from its higher diversity, the distribution of cloned sequences in this library was more even, with no individual phylotypes accounting for more than 3% of the community composition. Thus, as well as the two clone libraries differing in overall phylotype diversity and in general phylogenetic composition, they also differed in the specific groups of Bacteria that were most abundant within each.

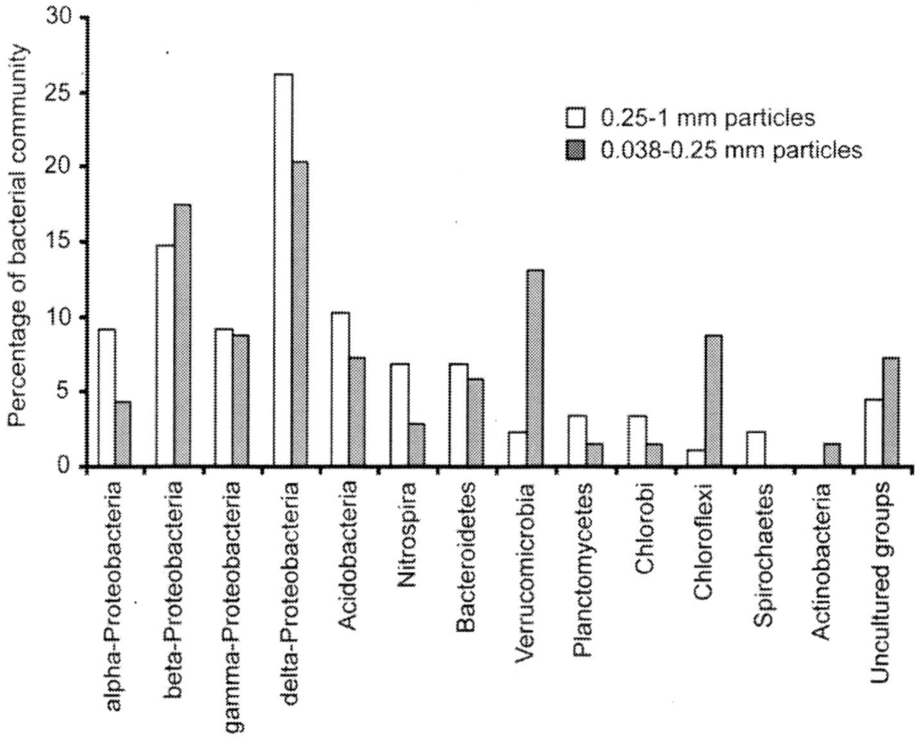

Figure 2. Frequency of major bacterial lineages in 16S rRNA clone libraries generated from PCR-amplified DNA extracted from two different sizes of fine particles (0.25-1mm diameter and 0.038-0.25 mm diameter) in wetland sediments. Representative of some bacterial lineages were found associated with both sizes of particles, but other groups are specific to just one type of particle, suggesting variation in bacterial community structure at fine scales.

Only one phylotype was found in both libraries. This is despite sequences being grouped into phylotypes by a fairly liberal 97% sequence similarity criterion, which would have led to increased overlap. This phylotype was affiliated with the Nitrosomonadales (beta-Proteobacteria) and was represented by just a single clone sampled in each library. Other than that phylotype, the two libraries contained no overlapping 16S rRNA genes, at least as apparent from sampling 96 clones within each. While it's possible that this may represent inadequate sampling, it strongly suggests that the different size ranges of particles harbor very different bacterial communities.

Differences between bacterioplankton and particle- or sediment-associated bacterial communities have been previously reported for aquatic environments [30, 102], but the influence of particle size on aquatic bacterial community structure or function has rarely been specifically addressed. This is in contrast to terrestrial ecosystems where soil particle size distributions (e.g. sand, silt, and clay) and their associations with organic matter are known to result in spatially heterogeneous environments [103]. This heterogeneity can result in differences in the bacterial populations associated with the different soil particle size fractions, and the influence of particle size on the structure of the bacterial community can be greater than that of larger scale treatments [104]. The results from this study clearly show differences in the bacterial phylotypes present in 16S rRNA clone libraries derived from DNA associated with 250-1000 μm and 38-250 μm diameter FPOM in wetland sediments. Sediment was separated into the two different size ranges by a wet sieving procedure, and because this involves washing the sediment with running water, unattached cells are likely to be removed. Thus, the differences in the bacterial phylotypes present in each size range are likely to reflect actual differences in the microbial communities attached to these types of particles.

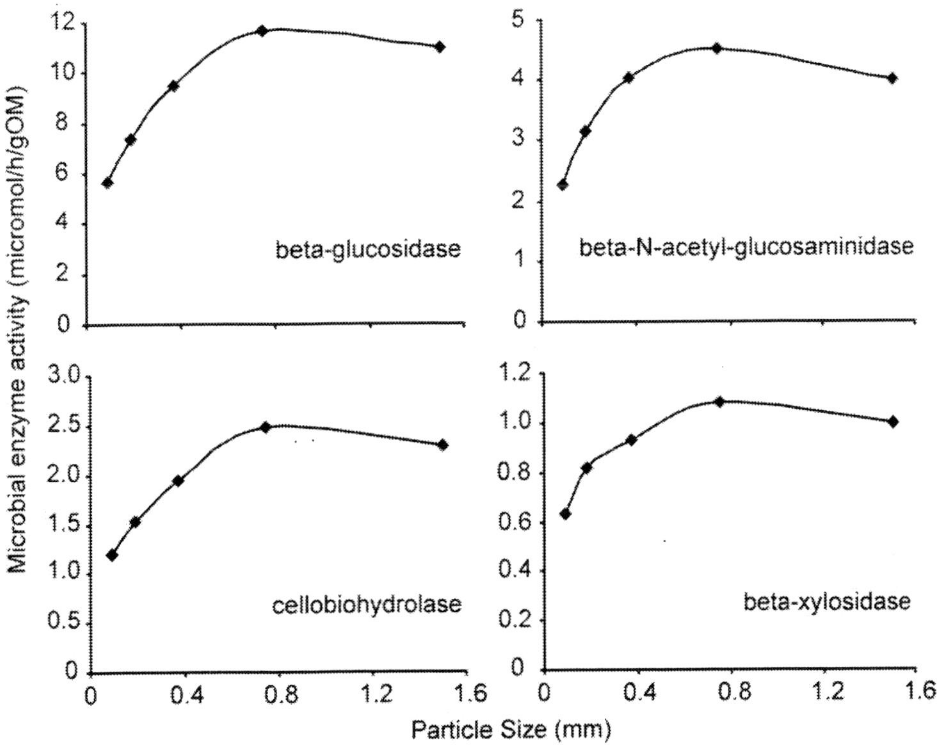

Figure 3. Changes in the activity of microbial extracellular enzymes associated with decreasing particle size in wetland sediments. Each point represents the mean of three sediment samples collected concurrently.

Functional differences in the microbial communities associated with different sizes of sediment particles were examined by assaying microbial extracellular enzyme activity. Enzymes such as cellobiohydrolase, beta-glucosidase, and beta-xylosidase are involved in the microbial degradation of lignocellulose and the activity of these enzymes has been linked to rates of POM decomposition [9, 105]. Sediment was collected from the same wetland site and sieved into five different sizes of particles of diameters 1000-2000 μm, 500-1000 μm, 250-500 μm, 125-250 μm, and 63-125 μm. Material in each size range was assayed for the activity of four extracellular enzymes (beta-glucosidase, cellobiohydrolase, beta-xylosidase, and beta-n-acetylglucosaminidase) using artificial p-nitrophenol linked substrates [97, 105].

Absorbance of the resulting end product was determined at 410 nm and converted to micromoles of substrate consumed using standard values [97].

Activity of all four enzymes was correlated (R values > 0.90 for each pair wise comparison) and declined on particles below 500 μm (Figure 3). This decline was consistent as particle size was further reduced, with the lowest activity of all enzymes being found associated with particles collected on the 63 μm sieve. The decline in activity on particles finer than 500 μm strongly suggests functional differences in the attached microbial community, and supports the differences in bacterial community structure reported above. Given that extracellular enzyme activity is related to rates of organic matter decomposition and nutrient cycling, these fine scale differences in microbial activity with particle size suggest that rates of lignocellulose degradation and other aspects of ecosystem function also differ over fine spatial scales.

FINE SCALE SPATIAL VARIATION IN MICROBIAL ACTIVITY ACROSS THE SURFACE OF DECOMPOSING LEAVES

Variation in microbial enzyme activity and community structure with different sizes of fine particulate organic matter is one example of how fine scale environmental variation affects aquatic microbial communities. A more direct measurement of spatial variation in microbial activity could be made by examining distribution patterns of enzyme activity across the surface of larger particles, such as decomposing leaves in aquatic systems. The surface of living leaves, or phyllosphere, supports variable microbial populations of bacteria and fungi and colonization of the phyllosphere is random and patchy [106]. However, no studies have looked at spatial variation in microbial activity on decomposing leaf-litter at fine scales.

In this study we examined the fine-scale variation in microbial extracellular enzyme activity across the surface of decomposing leaves in a small stream. As we stated previously in this chapter, the activities of lignocellulose-degrading enzymes are a reliable measure of rates of cellulose and lignin decomposition, as leaf-litter is composed primarily of lignin and cellulose, both of which require enzymatic degradation to be metabolized [9]. Here we measure the activities of two extracellular enzymes involved in cellulose decomposition (beta-glucosidase and cellobiohydrolase), and a combination of peroxidase and phenol oxidase activity as a broad measurement of enzymes involved in lignin decomposition.

Freshly fallen leaves of sycamore (*Platanus occidentalis*), northern red oak (*Quercus coccinea*), scarlet oak (*Quercus rubra*), and wild black cherry (*Prunus serotina*) were collected from Cypress Creek, a third order stream which flows through Holly Springs National Forest in northern Mississippi. Leaves were collected in the early winter, the period of greatest leaf fall at this site. Five replicate leaves of each species were collected from the surface of the streambed and refrigerated in water collected from the site until processing (1-3 days). On each day of processing, random leaves were chosen for assays of extracellular enzyme activity. Small circular discs (5 mm diameter) were bored at different points across each leaf (typically 15-35 points per leaf) and each disc was subsequently assayed for the activity of one microbial enzyme. Digital images of the remaining leaf (minus discs) were taken to identify the location of sample points for use in the subsequent spatial analysis.

Discs were assayed for enzymes using chromogenic artificial substrates. Beta-glucosidase and cellobiohydrolase activities were determined using the same p-nitrophenol linked substrates as for the wetland sediments study. Leaf discs were incubated with 300 μl substrate at room temperature for 2-4 hours, whereupon the reaction mixture was diluted with NaOH as described previously [97, 105]. Absorbance of the end product was determined at 405 nm [107] and converted to activity on an areal basis of leaf surface (i.e. activity as μmoles of substrate consumed per hour per area). Activities of phenol oxidase and peroxidase were combined into one measurement of overall oxidative enzyme activity using the artificial substrate L-3,4-dihydroxyphenylalanine (L-DOPA) and 0.015% hydrogen peroxide. Leaf discs were incubated in 300 μl of 5 mM L-DOPA with hydrogen peroxide for 2-4 hours and the absorbance of the diluted reaction mixture determined at 450 nm [108]. As with beta-glucosidase and cellobiohydrolase, oxidative enzyme activity was expressed as μmoles of substrate consumed per hour per area of leaf surface for each disc.

Activities of the microbial enzymes on the different discs taken from each leaf were used to calculate mean and standard deviation of activity for each enzyme and leaf species. These values were used to calculate the coefficient of variation (CV), a basic measurement of spatial variability [109, 110]. Spatial patterns in enzyme activity across the leaf surface were visualized by interpolation using the geostatistical software package SADA v.4.1.50 (Spatial Analysis and Decision Assistance, University of Tennessee).

In general, beta-glucosidase showed the highest activity of the three enzymes examined and cellobiohydrolase activity the lowest, regardless of leaf species (Figure 4). Oxidative enzyme activity and beta-glucosidase activity were similar on cherry and scarlet oak leaves (median oxidative activities of 0.04 and 0.05 $\mu mol/h/cm^2$, respectively, and median beta-glucosidase activities of 0.13 and 0.10 $\mu mol/h/cm^2$, respectively). These values were almost twice those seen on the other two leaf types (median oxidative activity was 0.03 $\mu mol/h/cm^2$ on both northern red oak and sycamore leaves, whereas median beta-glucosidase activity was 0.06 $\mu mol/h/cm^2$ for both). Median cellobiohydrolase activity was also similar for northern red oak and sycamore (0.01 $\mu mol/h/cm^2$) but differed substantially between scarlet oak (0.015 $\mu mol/h/cm^2$) and cherry (0.03 $\mu mol/h/cm^2$). These differences are likely because of a wide range of physical and chemical characteristics that differ between decomposing leaf types [111]. As would be expected based on their roles in cellulose degradation [112, 113], activities of beta-glucosidase and cellobiohydrolase were correlated, but this correlation ranged from weak ($R = 0.32$) on scarlet oak to moderately strong ($R = 0.79$) on cherry.

Variability in microbial activity across the surface of each leaf type was determined using a linear model fitted to the data comparing each sample point to the mean. Results of this model yielded significance values of $p < 0.001$ for all three enzymes on each leaf type, confirming a significant degree of spatial variability. CV values ranged from 29.5 to 43.8 for the different leaf types, and on any particular leaf were similar for the three different enzymes. Several-fold variation in activity between sample points across the surface of individual leaves was common for all three enzymes, and the range of activities reported for a given leaf was substantial (Figure 4).

Because the median enzyme activity and the range of activities were similar between replicates, a single northern red oak leaf was chosen for a more detailed representation of the variability in activity of each of the three microbial enzymes across the leaf surface (Data not shown). All three enzyme measurements showed fine scale variability with individual sample points differing in activity even over distances of a few centimeters. This variability was

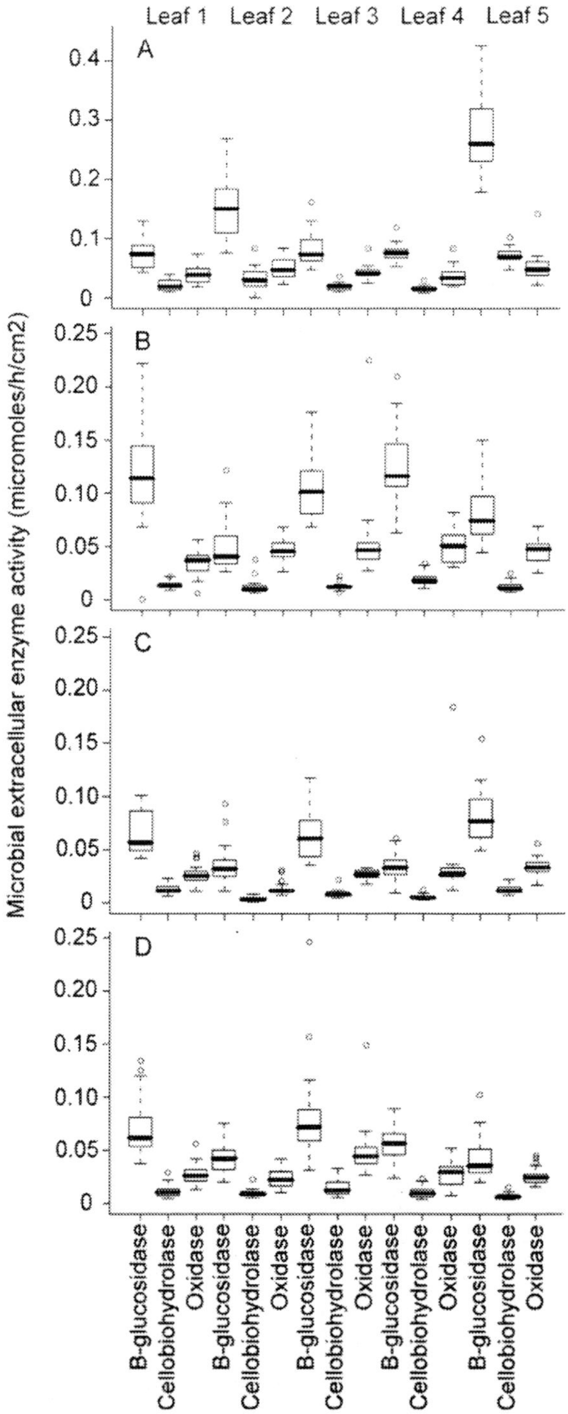

Figure 4. Statistical distributions for microbial extracellular enzyme activity sampled across of decomposing leaves. Five leaves of each species (A: cherry, B: scarlet oak, C: northern red oak, D: sycamore) were sampled for beta-glucosidase, cellobiohydrolase, and a combination of phenol oxidase and peroxidase activity. Solid lines for each leaf-enzyme combination indicate median values, with boxes indicating 25th and 75th quartiles, and whiskers representing 1.5x interquartile range. Most leaves showed high spatial variability in microbial activity even when outlying samples (open circles) were excluded.

typical of that seen in other leaves of northern red oak and in the other leaf types. Different spatial distributions of activity were visible for the three enzymes. Both beta-glucosidase and the combined oxidative enzymes showed distributions characterized by low background levels of activity with isolated "hot spots" of much higher enzyme activity. In contrast, cellobiohydrolase activity was intermediate on most of the discs sampled with intermittent points of both high and low activity. This finding is somewhat unusual given that activities of cellobiohydrolase and beta-glucosidase are more typically correlated [105], and the oxidative enzymes tend to show distributions different from those of the cellulases [105, 114]. However, most studies of microbial enzymatic activity sample at much larger scales than those examined here, and fine scale differences in environmental conditions across the leaf surface may be more important in influencing the distribution of some enzymes over others. Regardless, all three enzymes showed appreciable fine scale spatial heterogeneity in activity, even across the surface of a single leaf.

CONCLUSION

Aquatic microbial communities clearly function at fine scales and studies suggest that environmental heterogeneity at a level finer than that commonly examined in environmental microbiology can influence microbial community structure and function. This is apparent from planktonic communities, where differences in bacterial community composition and activity can occur between free-living cells and those associated with particles. Microorganisms in biofilms also show clear spatial structure, largely because of environmental gradients that form within the biofilm matrix. Similar gradients occur with depth in sediments, although other fine scale environmental variation in sediments are also apparent. A survey of microbial activity and community structure in wetland sediment showed that particle size plays an important role in determining the microorganisms present, with substantially different bacterial communities being found on fine particles less than 250 μm compared to those greater than that diameter. These structural changes are also reflected in changes in microbial enzyme activity as particle size decreases. Microbial enzyme activity itself can be used to assess fine scale spatial patterns, such as those that occur on the surface of decomposing leaves. Activities of enzymes involved in lignocellulose degradation show high spatial variability over individual leaves, again demonstrating that microbial communities vary both structurally and functionally over finer scales than are usually considered in environmental microbiology. The impact of this variation on broader processes is unknown, but fine scale environmental variation should be considered in future surveys assesing the microbial diversity and activity in aquatic environments in response to natural and anthropogenic forces.

REFERENCES

[1] Lindeman, R. (1942). The trophic-dynamic aspect of ecology. *Ecology, 23,* 399-417.
[2] Brar, S. K., Verma, M., Surampalli, R. Y., Misra, K., Tyagi, R. D., Meunier, N. And Blais, J. F. (2006). Bioremediation of hazardous wastes- a review. *Practice Periodical*

of Hazardous, Toxic, and Radioactive Waste Management, 10, 59-72.

[3] Amann, R. and Kühl, M. (1998). *In situ* methods for assessment of microorganisms and their activities. *Current Opinion in Microbiology, 1,* 352-358.

[4] Paerl, H. W. and Pinckney, J. L. (1996). A mini-review of microbial consortia: their roles in aquatic production and biogeochemical cycling. *Microbial Ecology, 31,* 225-247.

[5] Allen, T. F. H. and Hoekstra, T. W. (1992). *Toward a Unified Ecology*. New York, NY: Columbia University Press.

[6] O'Neill, R. V. (1989). Perspectives in Hierarchy and Scale. In J. Roughgarden, R. M. May, and S. A. Levin (Eds.), *Perspectives in Ecological Theory* (140-156). Princeton NJ: Princeton University Press.

[7] Krembs, C., Juhl, A. R., Long, R. A. and Azam, F. (1998). Nanoscale patchiness of bacteria in lake water studied with the Spatial Information Preservation method. *Limnology and Oceanography, 43,* 307-314.

[8] Moter, A. and Gobel, U. B. (2000). Fluorescence in situ hybridization (FISH) for direct visualization of microorganisms. *Journal of Microbiological Methods, 41,* 85-112.

[9] Sinsabaugh, R. L. (1994). Enzymic analysis of microbial pattern and process. *Biology and Fertility of Soils, 17,* 69-74.

[10] Santegoeds, C. M., Schramm, A. and de Beer, D. (1998). Microsensors as a tool to determine chemical microgradients and bacterial activity in wastewater biofilms and flocs. *Biodegradation, 9,* 159-167.

[11] Riemann, B. and Christofferson, K. (1993). Microbial trophodynamics in temperate lakes. *Aquatic Microbial Ecology, 7,* 69-100.

[12] Vannote, R. L., Minshall, G. W., Cummins, K. W., Sedell, J. R. and Cushing, C. E. (1980). The river continuum concept. *Canadian Journal of Fisheries and Aquatic Sciences, 37,* 130-137.

[13] Seymour, J. R., Mitchell, J. G., Pearson, L. and Waters, R. L. (2000). Heterogeneity in bacterioplankton abundance from 4.5 millimetre resolution sampling. *Aquatic Microbial Ecology, 22,* 143-153.

[14] Simon, M., Grossart, H. P., Schweitzer, B. and Ploug, H. (2002). Microbial ecology of organic aggregates in aquatic ecosystems. *Aquatic Microbial Ecology, 28,* 175-211.

[15] Cole, J. J. (1982). Interactions between bacteria and algae in aquatic ecosystems. *Annual Review of Ecology and Systematics, 13,* 291-314.

[16] Hahn, M. W., Moore, E. R. B. and Höfle, M. G. (2000). Role of microcolony formation in the protistan grazing defense of the aquatic bacterium *Pseudomonas* sp. MWH1. *Microbial Ecology, 39,* 175-185.

[17] Böckelmann, U., Manz, W., Neu, T. and Szewyzk, U. (2000). Characterization of the microbial community of lotic aggregates ('river snow') in the Elbe River of Germany by cultivation and molecular methods. *FEMS Microbiology Ecology, 33,* 157-170.

[18] Holliday, D. V., Donaghy, P. L., Greenlaw, C. F., McGehee, D. E., McManus, M. M., Sullivan, J. M. and Miksis, J. L. (2003). Advances in defining fine- and micro-scale pattern in marine plankton. *Aquatic Living Resources, 16,* 131-136.

[19] Ford, T. and Ryan, D. (1995). Toxic metals in aquatic ecosystems: a microbiological perspective. *Environmental Health Perspectives, 103,* 25-28.

[20] Moore, J. C., Berlow, E. L., Coleman, D. C., de Ruiter, P. C., Dong, Q., Hastings, A., Johnson, N. C., McMann, K. S., Melville, K., Morin, P. J., Nadelhoffer, K. Rosemond,

A. D., Post, D. M., Sabo, J. L., Scow, K. M., Vanni, M. J. and Wall, D. H. (2004). Detritus, trophic dynamics and biodiversity. *Ecology Letters, 7,* 584-600.

[21] Sherr, B. F., Sherr, E. B. and Hopkinson, C. S. (1988). Trophic interactions within pelagic microbial communities: indications of feedback regulation of carbon flow. *Hydrobiologia, 159,* 19-26.

[22] Alldredge, A. L., Passow, U. and Logan, B. E. (1993). The abundance and significance of a class of large, transparent organic particles in the ocean. *Deep Sea Research I, 40,* 1131-1140.

[23] Ploug, H. (2001). Small-scale oxygen fluxes and remineralization in sinking aggregates. *Limnology and Oceanography, 46,* 1624-1631.

[24] Ploug, H. and Passow, U. (2007). Direct measurement of diffusivity within diatom aggregates containing transparent exopolymer particles. *Limnology and Oceanography, 52,* 1-6.

[25] Smith, D. C., Simon, M., Alldredge, A. L. and Azam, F. (1992). Intense hydrolytic enzyme activity on marine aggregates and implications for rapid particle dissolution. *Nature, 359,* 139-141.

[26] Logan, B. and Wilkinson, D. B. (1990). Fractal geometry of marine snow and other biological aggregates. *Limnology and Oceanography, 35,* 130-136.

[27] Alldredge, A. L. (2000). Interstitial dissolved organic carbon (DOC) concentrations within sinking marine aggregates and their potential contribution to carbon flow. *Limnology and Oceanography, 45,* 1245-1253.

[28] Kiørbee, T., Ploug, H. and Thygesen, U. H. (2001). Fluid motion and solute distribution around sinking aggregates. I. Small-scale fluxes and heterogeneity of nutrients in the pelagic environment. *Marine Ecology Progress Series, 211,* 1-13.

[29] Crump, B. C., Baross, J. A. and Simenstad C. A. (1998). Dominance of particle-attached bacteria in the Columbia River estuary, USA. *Aquatic Microbial Ecology, 14,* 7-18.

[30] DeLong, E. F., Franks, D. G. and Alldredge, A. L. (1993). Phylogenetic diversity of aggregate-attached vs. free-living marine bacterial assemblages. *Limnology and Oceanography, 38,* 924-934.

[31] Schweitzer, B., Huber, I., Amann, R., Ludwig, W. and Simon, M. (2001). α- and β-*Proteobacteria* control the consumption and release of amino acids on lake snow aggregates. *Applied and Environmental Microbiology, 67,* 632-645.

[32] Grossart, H. P., Tang, K. W., Kiørbee, T. and Ploug, H. (2007). Comparison of cell-specific activity between free-living and attached bacteria using isolates and natural assemblages. *FEMS Microbiology Letters, 266,* 194-200.

[33] Grossart, H. P. and Ploug, H. (2000). Bacterial production and growth efficiencies: direct measurements on riverine aggregates. *Limnology and Oceanography, 45,* 436-445.

[34] Tang, K. W., Hutalle., K. M. L. and Grossart, H. P. (2006). Microbial abundance, composition and enzymatic activity during decomposition of copepod carcasses. *Aquatic Microbial Ecology, 45,* 219-227.

[35] Grossart, H. P., Kiørbee, T., Tang, K. W., Allgaier, M., Yam, E. M. and Ploug, H. (2006). Interactions between marine snow and heterotrophic bacteria: aggregate formation and microbial dynamics. *Aquatic Microbial Ecology, 42,* 19-26.

[36] Grossart, H. P., Kiørbee, T., Tang, K. and Ploug, H. (2003). Bacterial colonization of

particles: growth and interactions. *Applied and Environmental Microbiology, 69,* 3500-3509.

[37] Miki, T. and Yamamura, N. (2005). Theoretical model of interactions between particle-associated and free-living bacteria to predict functional composition and succession in bacterial communities. *Aquatic Microbial Ecology, 39,* 35-46.

[38] Mari, X., Kerros, M. E. And Weinbauer, M. G. (2007). Virus attachment to transparent exopolymeric particles along trophic gradients in the southwestern lagoon of New Caledonia. *Applied and Environmental Microbiology, 73,* 5245-5252.

[39] Diallo, M. S. and Savage, N. (2005). Nanoparticles and water quality. *Journal of Nanoparticle Research, 7,* 325-330.

[40] Branda, S. S., Vik, A., Friedman, L. and Kolter, R. (2005). Biofilms: the matrix revisited. *Trends in Microbiology, 13,* 20-26.

[41] Palková, Z. (2004). Multicellular microorganisms: laboratory versus nature. *EMBO Reports, 5,* 470-476.

[42] Donlan, R. M. (2002). Biofilms: microbial life on surfaces. *Emerging Infectious Diseases, 8,* 881-890.

[43] Douglas, S. and Beveridge, T. J. (1998). Mineral formation by bacteria in natural microbial communities. *FEMS Microbiology Ecology, 26,* 79-88.

[44] Jones, P. R., Cottrell, M. T., Kirchman, D. L. and Dexter, S. C. (2007). Bacterial community structure of biofilms on artificial surfaces in an estuary. *Microbial Ecology, 53,* 153-162.

[45] Bergey, E. A. (2006). Measuring the surface roughness of stream stones. *Hydrobiologia, 563,* 247-252.

[46] Purevdorj, B., Costerton, J. W. and Stoodley, P. (2002). Influence of hydrodynamics and cell signaling on the structure and behavior of *Psuedomonas aeruginosa* biofilms. *Applied and Environmental Microbiology, 68,* 4457-4464.

[47] Cardinale, B. J., Palmer, M. A., Swan, C. M., Brooks, S. and Poff, L. (2002). The influence of substrate heterogeneity on biofilm metabolism in a stream ecosystem. *Ecology, 83,* 412-422.

[48] Webster, N. S., Webb, R. I., Ridd, M. J., Hill, R. T. and Negri A. P. (2001). The effects of copper on the microbial community of a coral reef sponge. *Environmental Microbiology, 3,* 19-31.

[49] Sobolev, D. and Roden, E. E. (2002). Evidence for rapid microscale bacterial redox cycling of iron in circumneutral environments. *Antonie van Leeuwenhoek, 81,* 587-597.

[50] Okabe, S., Santegoeds, C. M., Watanabe, Y. and deBeer, D. (2002). Successional development of sulfate-reducing bacterial populations and their activities in a wastewater biofilm growing under microaerophilic conditions. *Applied and Environmental Microbiology, 68,* 1392-1402.

[51] Santegoeds, C. M., Ferdelman, T. G., Muyzer, G. and deBeer, D. (1998). Structural and functional dynamics of sulfate-reducing populations in bacterial biofilms. *Applied and Environmental Microbiology, 64,* 3731-3739.

[52] Schramm, A. (2003). In situ analysis of structure and activity of the nitrifying community in biofilms, aggregates, and sediments. *Geomicrobiology Journal, 20,* 313-333.

[53] Newman, D. K. and Banfield, J. F. (2002). Geomicrobiology: how molecular-scale interactions underpin biogeochemical systems. *Science, 296,* 1071-1077.

[54] Rowe, O. F., Sánchez-España, J., Hallberg, K. B. and Johnson, D. B. (2007). Microbial communities and geochemical dynamics in an extremely acidic, metal-rich stream at an abandoned sulfide mine (Huelva, Spain) underpinned by two functional primary production systems. *Environmental Microbiology, 9,* 1761-1771.

[55] Wimpenny, J., Manz, W. and Szewyzk, U. (2000). Heterogeneity in biofilms. *FEMS Microbiology Reviews, 24,* 661-671.

[56] Xu. K. D., Stewart, P. S., Xia, F., Huang, C. and McFeters, G. A. (1998). Spatial physiological heterogeneity in *Pseudomonas aeruginosa* biofilm is determined by oxygen availability. *Applied and Environmental Microbiology, 64,* 4035-4039.

[57] Whiteley, M., Bangera, M. G., Bumgarner, R. E., Parsek, M. R., Teitzel, G. M., Lory, S. and Greenberg, E. P. (2001). Gene expression in *Pseudomonas aeruginosa* biofilms. *Nature, 413,* 860-864.

[58] Horswill, A. R., Stoodley, P., Stewart, P. S. and Parsek, M. R. (2007). The effect of the chemical, biological, and physical environment on quorum sensing in structured microbial communities. *Analytical and Bioanalytical Chemistry, 387,* 371-380.

[59] Stoodley, P., deBeer, D. and Lewandowski, Z. (1994). Liquid flow in biofilm systems. *Applied and Environmental Microbiology, 60,* 2711-2716.

[60] Stewart, P. S. (2003). Diffusion in biofilms. *Journal of Bacteriology, 185,* 1485-1491.

[61] Francoeur, S. N., Wetzel, R. G. and Neely, R. K. (2001). New spatially explicit method for detecting extracellular protease activity in biofilms. *Applied and Environmental Microbiology, 67,* 4329-4334.

[62] Kalmbach, S., Manz, W. and Szewyk, U. (1997). Dynamics of biofilm formation in drinking water: phylogenetic affiliation and metabolic potential of single cells assessed by formazan reduction and in situ hybridization. *FEMS Microbiology Ecology, 22,* 265-279.

[63] Gadd, G. M. (2004). Microbial influence on metal mobility and application for bioremediation. *Geoderma, 122,* 109-119.

[64] Huws, S. A., McBain, A. J. and Gilbert, P. (2005). Protozoan grazing and its impact upon population dynamics in biofilm communities. *Journal of Applied Microbiology, 98,* 238-244.

[65] Lawrence, J. R. and Snyder, R. A. (1998). Feeding behaviour and grazing impacts of a *Euplotes* sp. on attached bacteria. *Canadian Journal of Microbiology, 44,* 623-629.

[66] Jackson, C. R., Churchill, P. F. and Roden, E. E. (2001). Successional changes in bacterial assemblage structure during epilithic biofilm development. *Ecology, 82,* 555-566.

[67] Jackson, C. R. (2003). Changes in community properties during microbial succession. *Oikos, 101,* 444-448.

[68] Hedin, L. O. (1990). Factors controlling sediment community respiration in woodland stream ecosystems. *Oikos, 57,* 94-105.

[69] Capone, D. G. and Kiene, R. P. (1988). Comparison of microbial dynamics in marine and freshwater sediments: Contrasts in anaerobic carbon catabolism. *Limnology and Oceanography, 33,* 725-749.

[70] Hutchison, N., Nagarkar, S., Aitchison, J. C. and Williams, G. A. (2006). Microspatial variation in marine biofilm abundance on intertidal rock surfaces. *Aquatic Microbial Ecology, 42,* 187-197.

[71] Tankere-Muller, S., Zhang, H., Davison, W., Finke, N., Larsen, O., Stahl, H. and Alud,

R. N. (2006). Fine scale remobilisation of Fe, Mn, Co, Ni, Cu and Cd in contaminated marine sediment. *Marine Chemistry, 106,* 192-207.

[72] Lavilla, I., Filgueiras, A. V., Valverde, F., Millos, J., Palanca, A. and Bendicho, C. (2006). Depth profile of trace elements in a sediment core of a high-altitude lake deposit at the Pyrenees, Spain. *Water, Air, and Soil Pollution, 172,* 273-293.

[73] Fischer, H., Sukhodolov, A., Wilczek, S. and Engelhardt, C. (2003). Effects of flow dynamics and sediment movement on microbial activity in a lowland river. *River Research and Applications, 19,* 473-482.

[74] Huettel, M., Røy, H., Precht, E. and Ehrenhauss, S. (2003). Hydrodynamical impact on biogeochemical processes in aquatic sediments. *Hydrobiologia, 494,* 231-236.

[75] Sinsabaugh, R. L. and Findlay, S. (1995). Microbial production, enzyme activity, and carbon turnover in surface sediments of the Hudson River estuary. *Microbial Ecology, 30,* 127-141.

[76] Liu, L., Fasheng, L., Xiong, D. and Song, C. (2006). Heavy metal contamination and their distribution in different size fractions of the surficial sediment of Haihe River, China. *Environmental Geology, 50,* 431-438.

[77] Webster, J. R. and Benfield, E.F. (1986). Vascular plant breakdown in freshwater ecosystems. *Annual Review of Ecology and Systematics, 17,* 567-594.

[78] Gooday, A. J., Turley, C. M. and Allen, J. A. (1990). Responses by benthic organisms to inputs of organic material to the ocean floor: A review (and discussion). *Philosophical Transactions of the Royal Society of London. Series A, Mathematical and Physical Sciences, 331,* 119-138.

[79] DeForest, D. K., Brix, K. V. and Adams, W. J. (2007). Assessing metal bioaccumulation in aquatic environments: the inverse relationship between bioaccumulation factors, trophic transfer factors and exposure concentration. *Aquatic Toxicology, 84,* 236-246.

[80] Wachinger, G., Fiedler, S., Zepp, K., Gattinger, A., Sommer, M. and Roth, K. (2000). Variability of soil methane production on the micro-scale: spatial association with hot spots of organic material and Archaeal populations. *Soil Biology and Biochemistry, 32,* 1121-1130.

[81] Bringmark, L. and Bringmark, E. (2001). Soil respiration in relation to small-scale patterns of lead and mercury in mor layers of southern Swedish forest sites. *Water, Air, and Soil Pollution: Focus, 1,* 395-408.

[82] Becker, J. M., Parkin, T., Nakatsu, C. H., Wilbur, J. D. and Konopka, A. (2006). Bacterial activity, community structure, and centimeter-scale spatial heterogeneity in contaminated soil. *Microbial Ecology, 51,* 220-231.

[83] Zorn, M. E., Lalonde, S. V., Gingras, M. K., Pemberton, S. G. and Konhauser, K. O. (2006). Microscale oxygen distribution in various invertebrate burrow walls. *Geobiology, 4,* 137-145.

[84] Stief, P. and de Beer, D. (2006). Probing the microenvironment of freshwater sediment macrofauna: Implications of deposit-feeding and bioirrigation for nitrogen cycling. *Limnology and Oceanography, 51,* 2538-2548.

[85] Stief, P., de Beer, D. and Neumann, D. (2002). Small-scale distribution of interstitial nitrite in freshwater sediment microcosms: the role of nitrate and oxygen availability, and sediment permeability. *Microbial Ecology, 43,* 367-378.

[86] Singh, B. K., Millard, P., Whiteley, A. S. and Murrell, J. C. (2004). Unraveling

rhizosphere-microbial interactions: opportunities and limitations. *Trends in Microbiology, 12,* 386-393.

[87] Hinsinger, P., Gobran, G. R., Gregory, P. J. and Wenzel, W. W. (2005). Rhizosphere geometry and heterogeneity arising from root-mediated physical and chemical processes. *New Phytologist, 168,* 293-303.

[88] Wotton, R. S. (1994). The classification of particulate and dissolved matter. In R. S. Wotton (Ed.), *The biology of particles in aquatic systems* (1-6). New York, NY: Lewis Publishers.

[89] Sinsabaugh, R. L. and Linkins, A. E. (1990). Enzymic and chemical analysis of particulate organic matter from a boreal river. *Freshwater Biology, 23*, 301-309.

[90] Ward, G. M., Ward, A, K., Dahm, C. N. and Aumen, N. G. (1994). Origin and formation of organic and inorganic particles in aquatic systems. In R. S. Wotton (Ed.), *The biology of particles in aquatic systems* (45-73). New York, NY: Lewis Publishers.

[91] Yeager, P. E. and Sinsabaugh, R. L. (1998). Microbial diversity along a sediment detrital particle size gradient. *Aquatic Ecology, 32*, 281-289.

[92] Lee, S. and Fuhrman, J. A. (1991). Spatial and temporal variation of natural bacterioplankton assemblages studies by total genomic DNA cross-hybridization. *Limnology and Oceanography, 36*, 1277-1287.

[93] Zhou, J., Bruns, M. A. and Tiedje, J. M. (1996). DNA recovery from soils of diverse composition. *Applied and Environmental Microbiology*, 62, 316-322.

[94] Jackson, C. R., Langner, H. W., Donahoe-Christiansen, J., Inskeep, W. P. and McDermott, T. R. (2001). Molecular analysis of microbial community structure in an arsenite-oxidizing acidic thermal spring. *Environmental Microbiology, 3*, 532-542.

[95] Jackson, C. R., Harper, J. P., Willoughby, D., Roden, E. E. and Churchill, P. F. (1997). A simple, efficient method for the separation of humic substances and DNA from environmental samples. *Applied and Environmental Microbiology, 63*, 4993-4995.

[96] Amann, R. I., Ludwig, W. and Schleifer, K. H. (1995). Phylogenetic identification and in situ detection of individual microbial cells without cultivation. *Microbiology Reviews, 59*, 143-169.

[97] Jackson, E. F., Echlin, H. L. and Jackson, C. R. (2006). Changes in the phyllosphere community of the resurrection fern, *Polypodium polypodioides*, associated with rainfall and wetting. *FEMS Microbiology Ecology, 58*, 236-246.

[98] Chao, A. (1987). Estimating the population size for capture-recapture data with unequal catchability. *Biometrics, 43*, 783-791.

[99] Kemp, P. F. and Aller, J. Y. (2004). Estimating prokaryotic diversity: When are 16S rDNA libraries large enough? *Limnology and Oceanography: Methods, 2*, 114-125.

[100] Ludwig, W., Strunk, O., Westram, R., Richter, L., Meier, H., Yadhukumar, Buchner, A., Lai, T., Steppi, S., Jobb, G., Förster, W., Brettske, I., Gerber,S., Ginhart, A. W., Gross, O., Grumann, S., Hermann, S., Jost, R., König, A., Liss, T., Lüßmann, R., May, M., Nonhoff, B., Reichel, B., Strehlow, R., Stamatakis, A., Stuckmann, N., Vilbig, A., Lenke, M., Ludwig, T., Bode, A. and Schleifer, K. H. (2004). ARB: a software environment for sequence data. *Nucleic Acids Research, 32*, 1363-1371.

[101] Nold, S. C. and Zwart, G. (1998). Patterns and governing forces in aquatic microbial communities. *Aquatic Ecology, 32*, 17-35.

[102] Besemer, K., Moeseneder, M. M., Arrieta, J. M., Herndl, G. J. and Peduzzi, P. (2005). Complexity of bacterial communities in a river-floodplain system (Danube, Austria).

Applied and Environmental Microbiology, 71, 609-620.

[103] Tisdall, J. M. and Oades, J. M. (1982). Organic matter and water stable aggregates in soils. *Journal of Soil Science, 33*, 141-163.

[104] Sessitsch, A., Weilharter, A., Gerzabek, M. H., Kirchmann, H. and Kandeler, E. (2001). Microbial population structures in soil particle size fractions of a long-term fertilizer field experiment. *Applied and Environmental Microbiology, 67*, 4215-4224.

[105] Jackson, C. R., Foreman, C. M. and Sinsabaugh, R. L. (1995). Microbial enzyme activities as indicators of organic matter processing rates in a Lake Erie coastal wetland. *Freshwater Biology, 34*, 329-342.

[106] Lindow, S. E. and Brandl, M. T. (2003). Microbiology of the phyllosphere. *Applied and Environmental Microbiology, 69,* 1875-1883.

[107] Blum, D. L., Kataeva, I. A., Li, X. L and Ljungdahl, L. G. (2000). Feruloyl esterase activity of the *Clostridium thermocellum* cellulosome can be attributed to previously unknown domains of XynY and XynZ. *Journal of Bacteriology, 182,* 1346-1351.

[108] Larson, J. L., Zak, D. R. and Sinsabaugh, R. L. (2002). Extracellular enzyme activity beneath temperate trees growing under elevated carbon dioxide and ozone. *Soil Science Society of America Journal, 66,* 1848-1856.

[109] Rossi, R. E., Mulla D. J., Journel, A. G. and Franz, E. H. (1992). Geostatistical tools for modelling and interpreting ecological spatial dependence. *Ecological Monographs, 62*, 277-314.

[110] Ettema, C. H. and Wardle, D. A. (2002). Spatial soil ecology. *Trends in Ecology and Evolution, 17*, 177-183.

[111] Cornelissen, J. H. C. (1996). An experimental comparison of leaf decomposition rates in a wide range of temperate plant species and types. *Journal of Ecology, 84,* 573-582.

[112] Divne, C., Stahlberg, J., Teeri, T. T. and Jones, T. A. (1998). High-resolution cryst structures reveal how a cellulose chain is bound in the 50 angstrom long tunnel of cellobiohydrolase I from *Trichoderma reesei. Journal of Molecular Biology, 275,* 309-325.

[113] Peshin, A. and Mathur, J. M. S. (1999). Purification and characterization of β-glucosidase from *Aspergillus niger* strain 322. *Letters in Applied Microbiology, 28,* 401-404.

[114] Alvarez, S. and Guerrero, M. C. (2000). Enzymatic activities associated with decomposition of particulate organic matter in two shallow ponds. *Soil Biology and Biochemistry, 32*, 1941-1951.

In: Aquatic Toxicology Research Focus
Editor: Elias P. Svensson, pp. 145-160
ISBN: 978-1-60456-192-0

Chapter 7

THE MICRONUCLEUS TEST FOR AQUATIC TOXICOLOGY

Ion Udroiu
Dipartimento di Biologia Animale e dell'Uomo, Università "La Sapienza", 00161 Roma, Italy

ABSTRACT

The growing concern about the presence of genotoxic substances in water bodies is leading to an increase of the attempts to refine the assays for the evaluation of genetic damage. Performing cytogenetic tests on ecologically relevant species allows carrying out assays to detect early effects of environmental exposure to genotoxicants. Moreover, these tests have also been widely used in laboratory studies to investigate the genotoxic properties of several compounds. Among these assays, one of the most popular is the micronucleus test, because of its simplicity, sensitivity and reliability. Micronuclei are nuclear remains produced during mitosis (or meiosis) when a chromosome fragment or an entire chromosome fails to migrate with one of the two daughter nuclei formed. Basically, this assay consists in the observation of the variations of the frequencies of micronucleated cells. The micronucleus test has been performed in many aquatic species, both invertebrate (worms, sea urchins, mussels) and vertebrate (mainly fish and amphibians, but also mammals). The samples can be taken from any active tissue or alternatively by induction of cellular proliferation. As this test has not yet been validated, special attention should be paid to protocols, experimental design, confounding factors and inter-species differences.

INTRODUCTION

In 1994, the OECD estimated that approximately 1,500 new chemicals are being added annually to the 100,000 already present in the natural environment. Along with the advancement of industrial technology, the characteristics of industrial discharge sources become more problematical and the toxicity of industrial wastewater can also become more complicated and serious. Natural and anthropogenic pollutants that contaminate the air and soil ultimately reach the aquatic environment, which represent the final sink for these substances. Accumulation and persistence of industrial effluents constitute a threat to biological life. It is well recognized that the disposal of pollutants into aqueous ecosystems can lead to their accumulation both in sediments and the benthic and pelagic food chains. Nonetheless, several fish and shellfish species represent a significant supply of protein and other nutrients in the diet of humans and human alimentation-destined animals. Therefore, several human populations are chronically exposed throughout their lifetime to toxicants present in both the water and aquatic food. For this reason, it is essential to know what effects, if any, water-borne pollutants produce on aquatic organisms.

Aquatic biomonitoring can be carried out detecting the concentrations of pollutants present in the tissues of marine and freshwater animals. This procedure is analogous to what in toxicology are called 'dose indicators'. In addition to dose indicators, effect indicators exist. These are biomarkers that reveal possible alterations of biological parameters. Whereas the use of dose indicators implies knowledge of what has to be searched, the use of effect indicators permits to detect the damage produced by substances not yet identified. Among these biomarkers, cytogenetic tests are particularly useful, as they allow detecting damage produced by numerous substances with probable carcinogenic properties [Tucker and Preston, 1996]. It is important to underline that Stahl [1991] indicated that about 30% of industrial waste waters are contaminated with genotoxic chemicals.

Genetic biomarkers detect changes in DNA or RNA from point mutations to changes in genetic diversity in populations [Belfiore and Anderson, 1998]. Genetic alterations in somatic cells can lead to decreased cell survival or transformation and can result in serious individual effects, like reproductive deficiencies and cancer [Shugart, 2000].

Among existing cytogenetic assays, the micronucleus test is frequently used to detect structural and numerical chromosomal aberrations induced by clastogenic and aneugenic agents. Until now, it proved to be a suitable tool both to evaluate the genotoxicity of xenobiotics in laboratory research and to assess the water quality during *in situ* studies.

MICRONUCLEI

Micronuclei are nuclear remnants, containing either a whole chromosome or a chromosome fragment. They are round or oval in shape, with a diameter usually ranging from 1/20 to 1/5 of the cell diameter, depending on the amount of chromatin included. By the way, some micronuclei are so small that they can be detected only by electron microscopy [Stich et al., 1990]. They reside in the cytoplasm, outside of the nucleus, to which they resemble in shape, structure and staining characteristics. These inclusions may be virtually found in any kind of cell, both in the somatic and the germinal ones. Being acentric, a chromatid fragment (produced by a clastogenic event) is not correctly segregated during anaphase. Therefore, it will not be included in one of the two nuclei of the daughter cells [Heddle and Carrano, 1977]. On the other hand, if the mitotic spindle suffers a damage (called an aneugenic one), a whole chromosome will be missegregated during karyocynesis. In this way, one of the daughter cells will bear a micronucleus alongside the nucleus. It may happen that a single cell contains more than a micronucleus, but the presence of three or more of these is interpreted as symptom of apoptosis.

Differently from those injuries detectable only in visible chromosomes (that is to say in cells undergoing mitosis), micronuclei are visible at any moment of the cell cycle.

The micronucleus test consists in the study of the variations of the frequencies of micronucleated cells. Usually, the comparison is done between two or more group of animals. This means, for example, comparing animals that received toxicants in laboratory, with others that did not, organisms from a polluted site with others from a protected area, or, at last, fish living downstream a source of pollution with those living upstream. On the other hand, this assay may be performed on cells sampled from the same group of animal at different moments.

As stated earlier, micronuclei can be found in any kind of cells, thus this test may be carried on with cells sampled from any proliferating tissue, such as kidney, spleen, gill epithelium and peripheral blood. Moreover, non-proliferating tissues like liver and caudal fin epithelium can also be used, provided cellular proliferation is induced. Beside these *in vivo* applications of the assay, also *ex vivo* methods can be adopted, that means working *in vitro* with cells sampled from the animals under study.

INVERTEBRATES

The micronucleus test has been widely applied on invertebrates, both in laboratory and *in situ*. Until now, samples have been taken from polychaetes, gastropods, sea urchins and, above all, mussels.

The nereid marine worm *Platynereis dumerilii* was used for the assessment of genotoxicity of municipal sewage effluents [Jha et al., 1997], tributyltin [Hagger et al., 2002] and benzo(*a*)pyrene [Jha et al., 1996]. The effects of colchicine were studied in *Pomatoceros lamarckii* [Wilson et al., 2002]. In a field study, Baršienė and Lovejoy [2000] found significant differences between the micronuclei frequencies of gastropods collected in contaminated and clean sites.

Hose [1985] analyzed micronuclei induction in sea urchin (*Strongylocentrotus purpuratus*) embryos exposed to benzo(*a*)pyrene, while Saotome et al. [1999] tested the effects of vinblastine sulphate, mitomycin C and 1-beta-D-arabinofuranosylcytosine on embryos of *Hemicentrotus pulcherrimus* and *Clypeaster japonicus*. It should be underlined that the use of embryonic assays offers the opportunity to screen simultaneously genotoxicity, teratogenicity and embryotoxicity.

Several environmental genotoxicity studies have been performed using bivalves from genus *Mytilus* [Scarpato et al., 1990; Brunetti et al., 1992; Dailianis et al., 2003; Izquierdo et al., 2003; Bolognesi et al., 2004; Kalpaxis et al., 2004; Hagger et al., 2005]. In most of the studies, higher frequencies of micronuclei have been detected in polluted areas, indicating the suitability of this test as a marker of cytogenetic damage in aquatic ecosystems.

In laboratory studies with mussels, micronuclei frequencies were found to increase in gill cells after treatment with mitomycin C, zinc chloride, colchicine [Majone et al., 1987; 1988; 1990], vincristine sulphate, benzo(*a*)pyrene [Burgeot et al., 1995]. At the same manner, significant increases have been found both in gill cells and haemocytes of mussels experimentally exposed to bisphenol A, diallylphtalate, tetrabromodifenyl ether [Baršienė et al., 2006a], benzo(*a*)pyrene [Venier et al., 1997; Siu et al., 2004] and dimethylbenzol(*a*)anthracene [Bolognesi et al., 1996].

The use of haemocytes is suitable for this assay as they are homogeneously exposed to toxicants and resulted sensitive to well-known genotoxicants. Until now however, little attention has been paid to the kinetics of the cell cycle and its influence upon micronucleation. Moreover, different studies adopted different sampling times. As a result, this test is not yet validated.

For the comparison of the micronuclei frequencies, it is important to discriminate the different sub-populations of haemocytes. In bivalves, for example they are comprised of agranular and granular cells [Cheng, 1981].

The use of bivalve gill cells is not as easy as haemocytes, being an invasive method and requiring enzymatic treatments followed by washes and incubation in different buffers to obtain single cell suspensions [Bolognesi et al., 1999].

Studies comparing micronuclei frequencies in haemocytes and gill epithelial cells of mussels showed that the latter tend to have higher baseline and induced frequencies [Mersch et al., 1996; Bolognesi et al., 1999; Dolcetti and Venier, 2002; Dailianis et al., 2003]. This may be due to the fact that the gill epithelium is the primary target for contaminants.

FISH

Undoubtedly, fish are the most used animals in the field of aquatic genotoxicology. They can be used to test the possible genotoxic properties of chemical and physical agents, as they seem to respond to xenobiotics in the same way as mammals. For example, Stegeman [1993] demonstrated that in fish hepatic activation of carcinogens is similar to the mammalian one. However, they have also been profitably used for biomonitoring purposes, both with feral and caged organisms.

Both laboratory research and *in situ* studies have involved several teleost species such as Cypriniformes [Al-Sabti, 1986; Kurihara et al., 1992; Ueda et al., 1992; Al-Sabti, 1994; Al-Sabti et al., 1994; de Lemos et al., 2001; Arkhipchuk and Garanko, 2005; Çavaş et al., 2005; Çavaş and Konen, 2007; Gül et al., 2007; Winter et al., 2007], Perciformes [Manna and Sadhukhan, 1986; Ieradi et al., 1996; Palhares and Grisolia, 2002; Farah et al., 2003; Gravato and Santos, 2003; Rodriguez-Cea et al., 2003; Arkhipchuk and Garanko, 2005; Van Ngan et al., 2007], Characiformes [Pantaleão et al., 2006], Anguilliformes [Rodriguez-Cea et al., 2003], Gadiformes [Baršienė et al., 2005; Baršienė et al., 2006b], Pleuronectiformes [Baršienė et al., 2005; Baršienė et al., 2006b; Bolognesi et al., 2006], Salmoniformes [Williams and Metcalfe, 1992; Schultz et al., 1993; Rao et al., 1997; Rodriguez-Cea et al., 2003] and Siluriformes [Metcalfe, 1988; Bahari et al., 1994; Ahmad et al., 2002; Ateeq et al., 2002].

Moreover, in two studies chondrichthyan [Borucinska et al., 2002] and chondrostean species [Bickham et al., 1998] have been used.

The micronucleus assay has been used to evaluate the genotoxic potential of aflatoxin B1, Aroclor 1254, benzidine [Al-Sabti, 1986], benzo(*a*)pyrene [Al-Sabti, 1986; Metcalfe, 1988; Gravato and Santos, 2003], bisphenol A [Baršienė et al., 2005; Bolognesi et al., 2006], pentachlorophenol [Ahmad et al., 2002; Farah et al., 2003], 2,4-dichlorophenoxyacetic acid

[Ateeq et al., 2002; Farah et al., 2003], tetrabromodiphenyl ether-47 [Baršienė et al., 2005; Bolognesi et al., 2006], diallyl phthalate, [Baršienė et al., 2005], dialkyl phthalate [Bolognesi et al., 2006], diethylnitrosamine [Williams and Metcalfe, 1992], ethylmethanesulphonate [Metcalfe, 1988; Williams and Metcalfe, 1992; Ahmad et al., 2002], butachlor [Farah et al., 2003], cadmium [Manna and Sadhukhan, 1986; Rodriguez-Cea et al., 2003; Arkhipchuk and Garanko, 2005; Çavaş et al., 2005], chromium [Al-Sabti et al., 1994; de Lemos et al., 2001], mercury, selenium [Al-Sabti, 1994], copper [Arkhipchuk and Garanko, 2005; Çavaş et al., 2005], chloral hydrate [Arkhipchuk and Garanko, 2005], mitomycin C [Ueda et al., 1992; Williams and Metcalfe, 1992; Bahari et al., 1994; Palhares and Grisolia, 2002; Winter et al., 2007], cyclophosphamide [Rodriguez-Cea et al., 2003; Winter et al., 2007], glyphosate [Çavaş and Konen, 2007], malathion [Gül et al., 2007], crude oil [Baršienė et al., 2006b] and X-rays [Manna and Sadhukhan, 1986; Kurihara et al., 1992; Schultz et al., 1993; Bahari et al., 1994].

Biomonitoring studies have been performed, both with marine [Borucinska et al., 2002; Van Ngan et al., 2007] and freshwater species [Ieradi et al., 1996; Bickham et al., 1998; Pantaleão et al., 2006], comparing the micronuclei frequencies of animals caught in polluted sites with those from preserved areas.

Several tissues have been chosen to carry out the piscine micronucleus test. Certainly, erythrocytes from peripheral blood are the most used [Al-Sabti, 1986; Metcalfe, 1988; Ueda et al., 1992; Schultz et al., 1993; Al-Sabti et al., 1994; Bahari et al., 1994; Bickham et al., 1998; de Lemos et al., 2001; Ateeq et al., 2002; Farah et al., 2003; Baršienė et al., 2005; Çavaş et al., 2005; Bolognesi et al., 2006; Çavaş and Konen, 2007; Gül et al., 2007; Van Ngan et al., 2007; Winter et al., 2007]. However, erythrocytes may also be obtained from the spleen [Winter et al., 2007] and cephalic kidney [Palhares and Grisolia, 2002; Rodriguez-Cea et al., 2003; Baršienė et al., 2005; Baršienė et al., 2006b], the latter being the main hemopoietical organ. Also gill epithelium cells are suitable for the application of the assay [Manna and Sadhukhan, 1986; Çavaş et al., 2005] and they proved to be the most sensitive type of cell to genotoxic insults [Ueda et al., 1992]. Moreover, the micronucleus test can be also performed on other tissues by stimulation of cell division. This has been obtained by damaging the edge of caudal fins [Arkhipchuk and Garanko, 2005]. In liver instead, cellular proliferation can be induced by exposure to allyl formate [Williams and Metcalfe, 1992; Rao et al., 1997; Çavaş et al., 2005]. The hepatic micronucleus test is particularly appropriate for those chemicals that require metabolic activation to become genotoxic.

Finally, the micronucleus assay has also been carried out *in vitro* with fin fibroblasts [Kurihara et al., 1992] and cytokinesis-blocked erythrocytes [Al-Sabti, 1994].

AMPHIBIANS

Amphibians, especially larval stages, have been widely used to test the genotoxic properties of several chemicals. In fact, in the mid 80s, Jaylet and his colleagues developed the application of the micronucleus test on peripheral erythrocytes of newt larvae [Siboulet et al., 1984], later known as the Jaylet Test. After its first development with larvae of *Pleurodeles waltl*, it has been largely applied to *Xenopus laevis* tadpoles and, in a fewer cases, with other species and/or with adult animals.

In laboratory studies, the Jaylet test (i.e. using larvae of *P. waltl*) was used to detect the genotoxic properties of X-rays, [Siboulet et al., 1984; Fernandez et al., 1993], benzo(*a*)pyrene [Siboulet et al., 1984; Jaylet et al., 1986a; Fernandez et al., 1989; Le Curiex et al., 1993; Djomo et al., 1995], naphthalene, anthracene, phenanthrene [Djomo et al., 1995], formaldehyde [Siboulet et al., 1984; Le Curiex et al., 1993], carbaryl, N-nitrosocarbaryl, aziridine, caffeine, [Siboulet et al., 1984], ethylmethanesulphonate, diethyl sulphate, N-ethyl-N'-nitro-N-nitrosoguanidine [Jaylet et al., 1986a], Aroclor 1254, butylated hydroxyanisole, phenobarbital, 12-O-tetradecanoyl-phorbol-13-acetate, acridine orange, ε-caprolactam, epichlorhydrin, N-ethyl-N-nitrosourea, hexamethylphosphoramide, pyrene, o-toluidine [Fernandez et al., 1989], cyclophosphamide [Fernandez et al., 1989; Le Curiex et al., 1993], 4-nitroquinoline 1-oxide [Le Curiex et al., 1993], monochloramine [Gauthier et al., 1989], captan [Mouchet et al., 2006a], sodium hypochlorite, [Gauthier et al., 1989; Le Curiex et al., 1993], potassium dichromate, 2-naphthylamine [Le Curiex et al., 1993], chlorinated propanones [Le Curiex et al., 1994], halogenated acetonitriles [Le Curiex et al., 1995a] halogenated acetic acid [Giller et al., 1997] and trihalomethanes [Le Curiex et al., 1995b].

As stated above, tadpoles of *Xenopus laevis* have also been largely used. In fact, they have been used to detect the levels of micronuclei induced by benzo(*a*)pyrene [Van Hummelen, et al., 1989; Mouchet et al., 2005a] ethylmethanesulphonate, methylmethanesulphonate [Mouchet et al., 2005a], cyclophosphamide, N-methyl-N-nitrosourea, the pyrethroid Fastac 10 EC [Rudek and Rozek, 1992], captan [Mouchet et al., 2006a] and cadmium [Mouchet et al., 2006b].

Moreover, larvae of *Ambystoma mexicanum* have been used to perform the micronucleus test after exposure to benzo(*a*)pyrene and ethylmethanesulphonate [Jaylet et al., 1986b]. At the same manner, tadpoles of *Rana catesbeiana* have been exposed to X-rays [Krauter et al., 1987], 2-acetylaminofluorene and 2-aminofluorene [Krauter, 1993], those of *Rana temporaria* to cyclophosphamide, N-methyl-N-nitrosourea and the pyrethroid Fastac 10 EC [Rudek and Rozek, 1992], those of *Rana limnocharis* to the pesticides imidacloprid and RH-5849 [Feng et al., 2004], tadpoles of *Odontophrynus americanus* to the pyrethroid cypermethrin [Cabagna et al., 2006] and those of *Hyla pulchella* to endosulfan [Lajmanovich et al., 2005]. The micronucleus test has also been performed on erythrocytes sampled from individuals belonging to *Triturus carnifex* exposed to mixtures of cadmium, zinc and copper [Capella et al., 1998].

It should also be added, that Zamora-Perez et al. [2004] determined the frequencies of micronucleated cells in the shed skin of *Ambystoma* exposed to cyclophosphamide and colchicine.

Concerning the field of biomonitoring properly said, the genotoxic properties of waste water, municipal sludge and dredged sediments have been studied exposing newt larvae [Gauthier et al., 1993] and *Xenopus* tadpoles [Chenon et al., 2003; Mouchet et al., 2005b; Mouchet et al., 2006c]. Moreover, ecogenotoxic monitoring has been conducted performing the micronucleus test on samples from adult belonging to *Rana ridibunda* captured *in situ* [Zhuleva and Dubinin, 1994; Matson et al., 2005a].

Finally, Tanzarella et al. [2006] studied the micronuclei frequencies in larval gill and adult erythrocytes and shed skin of *Triturus carnifex* inhabiting radon-rich ponds.

Therefore, the amphibian micronucleus assay proved to be a suitable tool both for testing chemical substances and biomonitoring studies. In particular, the larval stages are more sensitive (as it could be expected) to genotoxic insults than adults.

Reptiles

The micronucleus test has been performed in a very few cases on reptilian samples. However, these studies mainly deal with biomonitoring of the aquatic environment. Apart from the works of Zúñiga-González et al. [2000], who studied the baseline values of captive turtles and crocodiles, this assay has been applied on blood samples of *Emys orbicularis* and *Mauremys caspica*, showing a correlation between the frequencies of micronucleated erythrocytes and tissue concentrations of some pollutants [Swartz et al., 2003; Matson et al., 2005b].

Mammals

Until now, the micronucleus test has been widely used on samples of several mammalian species, both *in vivo* and *in vitro*. Actually, it was first developed on murine bone marrow erythrocytes [Matter and Schmid, 1971]. Since mammalian erythrocytes are anucleated, micronuclei can be easily detected, making this kind of cells the most used for conducting the test. In fact, they were first discovered in mammalian erythrocytes [Howell, 1891; Jolly, 1905] and are well-known in haematology as Howell-Jolly bodies.

Regarding the field of aquatic toxicology, this assay has been performed *in vivo* and *ex vivo*, using samples from marine species, but also exposing laboratory animals or cultures of mammalian cells to aquatic pollutants.

The only application of the micronucleus test on marine mammals erythrocytes is the recent work of Zamora-Perez et al. [2006]. It detected significant differences in the frequencies of micronucleated erythrocytes sampled from bottlenose dolphins (*Tursiops truncatus*) from different dolphinariums. However, it should be kept in mind for future studies that the application of this assay on circulating erythrocytes is not feasible in every mammalian species, as the spleen may remove the micronuclei [Udroiu, 2006a].

Gauthier et al. [1999a] used the cytokinesis block micronucleus (CBMN) test to study the baseline values of cetaceans (*Delphinapterus leucas* and *Tursiops truncatus*) and seals (*Halichoerus grypus* and *Phoca groenlandicus*) from relatively healthy wild populations.

The same group studied also the induction of micronuclei in skin fibroblasts sampled from beluga whales and exposed to organochlorine compounds [Gauthier et al., 1999b] and mercury [Gauthier et al., 1998].

Finally, it should be added that also the use of non-aquatic mammalian cultured cells may be useful for evaluating the genotoxicity of water samples from rivers, leachates, etc.

Protocols and Experimental Design

The species adopted in the experiment should be chosen considering also its karyotype. Many teleosts, for example, are characterized by an elevated number of small chromosomes. For these species, it is likely that micronuclei formed after a clastogenic event will be very small in size, some of them being not observable under light microscopy. For that reason, species with fewer but larger chromosomes, such as *Nothobranchius rachowi* (2n = 16),

Sphariclhys osphoromoides (2n = 16), *Aphyosemion christyi* (2n = 18), *Galaxias maculates* (2n = 22), *Umbra limi* (2n = 22) and *Umbra pygmaea* (2n = 22) are recommended [Kligerman, 1982]. Otherwise, the detection of very small micronuclei can be assured by fluorescent staining.

Knowledge of species-specific metabolic activation of genotoxic substances is also important.

The use of embryos or larval stages requires more attention in handling the animals and obtaining the samples, but it is characterized by a greater sensitiveness to genotoxic agents.

Once the species has been chosen, it should be assured that factors like age, sex, temperature and diet are the same in the exposed and the control groups. In fish, for example, inadequate feeding determines a lengthening (often dramatic) of the erythrocytes lifespan [Soldatov, 1995]. As the majority of animals used in aquatic toxicology are poikilotherms, temperature measurement and control is very important. In fact, the micronucleus assay needs that the cell population undergoes mitosis and the cell cycle depends on the water temperature.

The micronucleus test can be performed either *in situ*, comparing the frequencies of animals caught or caged in different sites, or in laboratory. As this assay is not yet validated, if a new chemical or mixture is studied, then a positive control group, besides the exposed one and the negative controls, should be formed. For this purpose, well-known mutagens, such as ethylmethanesulphonate, cyclophosphamide, colchicines or X-rays can be used.

In laboratories, animals should acclimatise for 2 weeks before the start of the experiment. Exposure may be achieved by a single dose, repeated doses or be continuous (i.e., the xenobiotic is continuously present in the water). Sampling time depends on the species and tissue used. In fish, a peak in micronucleated cells occurs 1-2 days after exposure, but in peripheral blood it takes place after 2 or 3 days.

While micronucleated cells from gills, liver, skin and the hemopoietic organs (spleen, kidney or bone marrow, depending on the species) reflect genotoxic damage which occurred during the last cell cycle, micronucleated erythrocytes from peripheral circulation are informative of events that occurred in a time equal to the lifespan of the circulating erythrocytes. Therefore, repeated blood sampling may be very useful in laboratory experiments. However, if blood from laboratory animals is sampled too frequently, erythropoiesis is stimulated, resulting in a dilution of the micronucleated erythrocytes previously formed, with a lowering of their frequency [Udroiu, 2006b].

After sampling, slides should be coded and "blindly" scored. As the presence of more scorers could be a source of uncontrolled variability, the slides should be scored by the same analyst. On the slide, the area to be scored should be chosen at low magnification on the basis of its quality. Subsequently, scoring of micronucleated cells frequencies should be performed at 1000× magnification. If the cells under exam are comprised of different sub-populations, these should be counted separately. For red blood cells, for examples, the ratio between immature and mature erythrocytes is also a useful index to assess cytotoxicity.

It is recommended to score at least 2000 cells per animal. However, many species show a very low frequency of spontaneous micronuclei. If weak genotoxicants are tested, the detection of a micronucleated cell may be a very rare event, weakening the power of statistical analysis. In this case, a higher number of cells per animal should be scored. This can be more easily achieved with the use of automated scoring devices [Ahmad et al., 2002].

Conclusion

In the last years, there have been regulatory developments which have suggested the genotoxicity testing of chemicals and effluents before their discharge into the aquatic environment. The US Environmental Protection Agency has recommended adding mutagenicity testing protocols to the existing set of chemical techniques adopted to detect the presence of specific chemicals or effluents [Houk, 1992]. Moreover, the International Standards Organisation (ISO) recommended a standard bacterial method for testing the genotoxicity of wastewater. It should be underlined that while *in vitro* or bacterial test systems may define the intrinsic genotoxicity of a substance, ecotoxicological risk assessments should also consider the environmentally realistic routes of exposure, the effects of metabolism and DNA repair efficiency.

Investigators became interested in the micronucleus test, as it had the potential of avoiding the inconveniences of other cytogenetic assays such as the chromosome aberration and the sister chromatid exchange tests. The latter techniques can effectively detect the genotoxic properties of chemical pollutants, but these tests are time consuming and do not seem to be very effective for the many species which have relatively large numbers of small chromosomes. The micronucleus assay proved to be sensitive and easy to perform, particularly with blood samples.

Although the validation of the micronucleus test in aquatic organisms seems not very close, its importance for evaluating the water quality makes its refining urgent. Future studies should elucidate the role of factors, such as temperature, age, sex, reproductive status and erythropoiesis.

References

Ahmad, W; Ali, MN; Farah, MA; Ateeq, B. Computerized automated morphometric assay including frequency estimation of pentachlorophenol induced nuclear anomalies (micronucleus) in catfish *Heteropneustes fossilis*. Chromosoma, 2002 110, 570-574.

Al-Sabti, K. Comparative micronucleated erythrocyte cell induction in three cyprinids by five carcinogenic-mutagenic chemicals. *Cytobios*, 1986 47, 147-154.

Al-Sabti, K. Micronuclei induced by selenium, mercury, methylmercury and their mixtures in binucleated blocked fish erythrocyte cells. *Mutat. Res.*, 1994 320, 157-163.

Al-Sabti, K; Franko, M; Andrijani, B; Knez, S; Stegnar, P. Chromium induced micronuclei in fish. *J. Appl. Toxicol.,* 1994 14, 333-336.

Arkhipchuk, VV; Garanko NN. Using the nucleolar biomarker and the micronucleus test on in vivo fish fin cells. *Ecotoxicol. Environ. Saf.*, 2005 62, 42-52.

Ateeq, B; Farah, MA; Ali, MN; Ahmad W. Induction of micronuclei and erythrocyte alterations in the catfish *Clarias batrachus* by 2,4-dichlorophenoxyacetic acid and butachlor. *Mutat. Res.*, 2002 518, 135-144.

Bahari, IB; Noor, FM; Daud, NM. Micronucleated erythrocytes as an assay to assess actions by physical and chemical genotoxic agents in *Clarias gariepinus*. *Mutat. Res.*, 1994 313, 1-5.

Baršienė, J; Dedonytė, V; Rybakovas, A; Andreikėnaitė, L; Andersen, OK. Induction of micronuclei in Atlantic cod (*Gadus morhua*) and turbot (*Scophthalmus maximus*) after treatment with bisphenol A, diallyl phthalate and tetrabromodiphenyl ether-47. *Ekologija*, 2005 4, 1-7.

Baršienė, J; Syvokiene, J; Bjornstad, A. Induction of micronuclei and other nuclear abnormalities in mussels exposed to bisphenol A, diallyl phthalate and tetrabromodiphenyl ether-47. *Aquat. Toxicol.*, 2006a 78S, S105-108.

Baršienė, J; Dedonytė, V; Rybakovas, A; Andreikėnaitė, L; Andersen, OK. Investigation of micronuclei and other nuclear abnormalities in peripheral blood and kidney of marine fish treated with crude oil. *Aquat. Toxicol.*, 2006b 78S, S99-104.

Belfiore, NM; Anderson, SL. Genetic patterns as a tool for monitoring and assessment of environmental impacts: The example of genetic Ecotoxicology. *Enviorn. Monit. Assess.*, 1998 51, 465-479.

Bickham, JW; Rowe, GT; Palatnikov, G; Mekhtiev, A; Mekhtiev, M; Kasimov, RY; Hauschultz, DW; Wickliffe, JK; Rogers, WJ. Acute and genotoxic effects of Baku Harbor sediment on Russian sturgeon, *Acipenser guildensteidti*. *Bull. Environ. Contam. Toxicol.*, 1998 61, 512-518.

Bolognesi, C; Rabboni, R; Roggieri, P. Genotoxicity biomarkers in *M. galloprovincialis* as indicators of marine pollutants. *Comp. Biochem. Physiol.*, 1996 113C, 319-323.

Bolognesi, C; Landini, E; Roggieri, P; Fabbri, R; Viarengo, A. Genotoxicity biomarkers in the assessment of heavy metal effects in mussels: experimental studies. *Environ. Mol. Mutagen,* 1999 33, 287-292.

Bolognesi, C; Frenzilli, G; Lasagna, C; Perrone, E; Roggieri, P. Genotoxicity biomarkers in *Mytilus galloprovincialis*: wild versus caged mussels. *Mutat. Res.*, 2004 552, 153-162.

Bolognesi, C; Perrone, E; Roggieri, P; Pampanin, DM; Sciutto, A. Assessment of micronuclei induction in peripheral erythrocytes of fish exposed to xenobiotics under controlled conditions. *Aquat. Toxicol.*, 2006 78, S93-S98.

Borucinska, JD; Czachorowska, A; Mookherjea, S; Habib, M. Micronuclei in peripheral red blood cells of *Mustelus canis* and *Squalus acanthias* representing two orders of sharks from the Long Island sound and coastal waters of the Northwestern Atlantic. Proc New England Estuarine Research Society / Long Island Sound Research Conf., Groton. CT, USA, 2002.

Brunetti, R; Gabriele, M; Valerio, P; Fumagalli, O. The micronucleus test: Temporal pattern of base-line frequency in *Mytilus galloprovincialis* . *Mar. Ecol. Prog. Ser.*, 1992 83, 75-78.

Burgeot, T; His, E; Galgani, F. The micronucleus assay in *Crassostrea gigas* for the detection of seawater genotoxicity. *Mutat. Res.*, 1995 342, 125-140.

Cabagna, MC; Lajmanovich, RC; Peltzer, PM; Attademo, AM; Ale E. Induction of micronuclei in tadpoles of *Odontophrynus americanus* (Amphibia: Leptodactylidae) by the pyrethroid insecticide cypermethrin. *Toxicol. Environ. Chem.*, 2006 88, 729-737.

Capella, S; Peyrot, A; Bigatti, MP; DePascalis, MG; Falsaperla, MG; Maurino, M; Vellano, C. Genotoxicity of heavy metals in the newt (*Triturus carnifex*): micronucleus test in erythrocytes. *Acc. Sc. Torino Atti. Sc. Fis.*, 1998 132, 1-10.

Çavaş, T; Garanko, NN; Arkhipchuk, VV. Induction of micronuclei and binuclei in blood, gill and liver cells of fishes subchronically exposed to cadmium chloride and copper sulphate. *Food Chem. Toxicol.,* 2005 43, 569-574.

Çavaş, T; Konen, S. Detection of cytogenetic and DNA damage in peripheral erythrocytes of goldfish (*Carassius auratus*) exposed to a glyphosate formulation using the micronucleus test and the comet assay. *Mutagenesis,* 2007, in press.

Cheng, TC. Bivalves. In: Ratcliffe, NA, Rowley, AF, editors. *Invertebrate Blood Cells.* London : Academic Press; 1981; Vol 1; 231-300.

Chenon, P; Gauthier, L; Loubieres, P; Severac, A; Delpoux, M. Evaluation of the genotoxic and teratogenic potential of a municipal sludge and sludge-amended soil using the amphibian *Xenopus laevis* and the tobacco: *Nicotiana tabacum* L. var. xanthi Dulieu. *Sci. Total Environ.*, 2003 301, 139-150.

Dailianis, S; Domouhtsidou, GP; Raftopoulou, E; Kaloyianni, M; Dimitriadis, VK. Evaluation of neutral red retention assay, micronucleus test, acetylcholinesterase activity and a signal transduction molecule (cAMP) in tissues of *Mytilus galloprovincialis* (L.), in pollution monitoring. *Mar. Environ .Res.*, 2003 56, 443-470.

de Lemos, CT; Rodel, PM; Terra, NR; Erdtmann, B. Evaluation of basal micronucleus frequency and hexavalent chromium effects in fish erythrocytes. *Environ. Toxicol.Chem.,* 2001 20, 1320-1324.

Djomo, JE; Ferrier, V; Gauthier, L; Zoll-Moreux, C; Marty, J. Amphibian micronucleus test in vivo: evaluation of the genotoxicity of some major polycyclic aromatic hydrocarbons found in a crude oil. *Mutagenesis*, 1995 10, 223-226.

Dolcetti, L; Venier, P. Susceptibility to genetic damage and cell types in Mediterranean mussels. *Mar. Environ. Res.*, 2002 54, 487-491.

Farah, MA; Ateeq, B; Ali, MN; Ahmad, W. Evaluation of genotoxicity of PCP and 2,4-D by micronucleus test in freshwater fish *Channa punctatus. Ecotoxicol. Environ. Saf.,* 2003 54, 25-29.

Feng, S; Kong, Z; Wang, X; Zhao, L; Peng, P. Acute toxicity and genotoxicity of two novel pesticides on amphibian, *Rana* N. Hallowell. *Chemosphere,* 2004 56, 457-463.

Fernandez, M; Gauthier, L; Jaylet, A. Use of newt larvae for in vivo genotoxicity testing of water: results on 19 compounds evaluated by the micronucleus test. *Mutagenesis,* 1989 4, 17-26.

Fernandez, M; L'Haridon, J; Gauthier, L; Zoll-Moreux, C. Amphibian micronucleus test(s): a simple and reliable method for evaluating in vivo genotoxic effects of freshwater pollutants and radiations. Initial assessment. *Mutat. Res.*, 1993 292, 83-99.

Gauthier, JM; Dubeau, H; Rassart, E. Mercury induced micronuclei in skin fibroblasts of beluga whales. *Environ. Toxicol. Chem.*, 1998 17, 2487-2493.

Gauthier, JM; Dubeau, H; Rassart, E; Jarman, WM; Wells RS. Biomarkers of DNA damage in marine mammals. *Mutat. Res.*, 1999a 444, 427-439.

Gauthier, JM; Dubeau, H; Rassart, E. Induction of micronuclei in vitro by organochlorine compounds in beluga whale skin fibroblasts. *Mutat. Res.*, 1999b 439, 87-95.

Gauthier, L; Levi, Y; Jaylet, A. Evaluation of the clastogenicity of water treated with sodium hypochlorite or monochloramine using a micronucleus test in newt larvae (*Pleurodeles waltl*). *Mutagenesis*, 1989 4, 170-173.

Gauthier, L; Van der Gaag, MA; L'Haridon, J; Ferrier, V; Fernandez, M. In vivo detection of waste water and industrial effluent genotoxicity: use of the Newt Micronucleus Test (Jaylet Test). *Sci. Total Environ.*, 1993 138, 249-269.

Giller, S; Le Curieux, F; Erb, F; Marzin, D. Comparative genotoxicity of halogenated acetic acids found in drinking water. *Mutagenesis*, 1997 12, 321-328.

Gül, S; Nur, G; Kaya, TÖ; Kamber, U; Gürdegin, B. Detection of micronuclei in peripheral erythrocytes of *Orthrias angorae* (Steindachner, 1897) exposed to malathion. *Fresenius Environ .Bull.*, 2007 16, 472-476.

Gravato, C; Santos, MA. Genotoxicity biomarkers' association with B(*a*)P biotransformation in *Dicentrarchus labrax* L. *Ecotoxicol. Environ. Saf.*, 2003 55, 352-358.

Hagger, JA; Fisher, AS; Hill, SJ; Depledge, MH; Jha, AN. Genotoxic, cytotoxic and ontogenetic effects of tri-n-butyltin on the marine worm, *Platynereis dumerilii* (Polychaeta: Nereidae). *Aquat. Toxicol.*, 2002 57, 243-255.

Hagger JA, Atienzar FA, Jha AN. Genotoxic, cytotoxic, developmental and survival effects of tritiated water in the early life stages of the marine mollusc, *Mytilus edulis*. *Aquat. Toxicol.*, 2005 74, 205-217.

Heddle, JA; Carrano, AV. The DNA content of micronuclei induced in mouse bone marrow by gamma-irradiation: evidence that micronuclei arise from acentric chromosomal fragments. *Mutat. Res.*, 1977 44, 63-69.

Hose, JE. Potential uses of sea urchin embryos for identifying toxic chemicals: description of a bioassay incorporating cytologic, cytogenetic and embryologic endpoints. *J. Appl. Toxicol.*, 1985 5, 245-254.

Houk, VS. The genotoxicity of industrial wastes and effluents. *Mutat. Res.*, 1992 277, 91-138.

Howell, WH. The life-history of the formed elements. of the blood, expecially of the red blood corpuscles. *J. Morphol.*, 1891 4, 57-116.

Ieradi, LA; Meucci, F; Giucca, F; Ciccotti, E; Cardarelli, E; Grossi, R; Campanella, L. Mutagenicity test and heavy metals in fish from Tiber river. *J. Ecol. Chem.*, 1996 5, 287-291.

Izquierdo, JI; Machado, G; Ayllon, F; d'Amico, VL; Bala, LO; Vallarino, E; Elias, R; Garcia-Vazquez E. Assessing pollution in coastal ecosystems: a preliminary survey using the micronucleus test in the mussel *Mytilus edulis*. *Ecotoxicol. Environ. Saf.*, 2003 55, 24-29.

Jaylet, A; Deparis, P; Ferrier, V; Grinfeld, S; Siboulet, R. A new micronucleus test using peripheral blood erythrocytes of the newt *Pleurodeles waltl* to detect mutagens in fresh-water pollution. *Mutat. Res.*, 1986a 164, 245-257.

Jaylet, A; Deparis, P; Gaschignard, D. Induction of micronuclei in peripheral erythrocytes of axolotl larvae following in vivo exposure to mutagenic agents. *Mutagenesis*, 1986b 1, 211-215.

Jha AN, Hutchinson TH, Mackay JM, Elliott BM, Dixon DR. Development of an in vivo genotoxicity assay using the marine worm *Platynereis dumerilii* (Polychaeta: Nereidae). *Mutat. Res.*, 1996 359, 141-150.

Jha AN, Hutchinson TH, Mackay JM, Elliott BM, Dixons DR. Evaluation of the genotoxicity of municipal sewage effluent using the marine worm *Platynereis dumerilii* (Polychaeta: Nereidae). *Mutat. Res.*, 1997 391, 179-188.

Jolly, J. Sur l'evolution des globules rouges dans le sans des embryons de mammiferes. C R Séanc Soc Biol, 1905 58, 593-595.

Kalpaxis, DL; Theos, C; Xaplanteri, MA; Dinos, GP; Catsiki, AV; Leotsinidis, M. Biomonitoring of Gulf of Patras, N. Peloponnesus, Greece. Application of a biomarker suite including evaluation of translation efficiency in *Mytilus galloprovincialis* cells. *Environ. Res.*, 2004 94, 211-220.

Kligerman, AD. Fishes as a biological detector of the effects of genotoxic agents. In: Heddle JA, editor. Mutagenicity: New Horizons in Genetic Toxicology. New York: Academic Press; 1982; 435-453.

Krauter, PW. Micronucleus incidence and hematological effects in bullfrog tadpoles (*Rana catesbeiana*) exposed to 2-acetylaminofluorene and 2-aminofluorene. *Arch. Environ. Contam. Toxicol.*, 1993 24, 487-493.

Krauter, PW; Anderson, SL; Harrison, FL. Radiation-induced micronuclei in peripheral erythrocytes of *Rana catesbeiana*: an aquatic animal model for in vivo genotoxicity studies. *Environ. Mol. Mutagen.*, 1987 10, 285-296.

Kurihara, Y; Rienkjkarn, M; Etoh H. Cytogenetic adaptive response of cultured fish cells to low doses of X-rays. *J. Radiat. Res.*, 1992 33, 267-274.

Lajmanovich, RC; Cabagna, M; Peltzer, PM; Stringhini, GA; Attademo, AM. Micronucleus induction in erythrocytes of the *Hyla pulchella* tadpoles (Amphibia: Hylidae) exposed to insecticide endosulfan. *Mutat. Res.*, 2005 587, 67-72.

Le Curieux, F; Marzin, D; Erb, F. Comparison of three short-term assays: results on seven chemicals. Potential contribution to the control of water genotoxicity. *Mutat. Res.*, 1993 319, 223-236.

Le Curieux, F; Marzin, D; Erb, F. Study of the genotoxic activity of five chlorinated propanones using the SOS chromotest, the Ames-fluctuation test and the newt micronucleus test. *Mutat. Res.*, 1994 341, 1-15.

Le Curieux, F; Giller, S; Gauthier, L; Erb, F; Marzin, D. Study of the genotoxic activity of six halogenated acetonitriles, using the SOS chromotest, the Ames-fluctuation test and the newt micronucleus test. *Mutat. Res.*, 1995a 341, 289-302.

Le Curieux, F; Gauthier, L; Erb, F; Marzin, D. Use of the SOS chromotest, the Ames-fluctuation test and the newt micronucleus test to study the genotoxicity of four trihalomethanes. *Mutagenesis,* 1995b 10, 333-341.

Manna, GK; Sadhukhan, A. Use of cells of gill and kidney of Tilapia fish in micronucleus test (MNT). *Current Science*, 1986 55, 498-501.

Matson, CW; Palatnikov, G; McDonald, TJ; Autenrieth, RL; Donnelly, KC; Anderson, TA; Canas, JE; Islamzadeh, A; Bickham, JW. Patterns of genotoxicity and contaminant exposure: evidence of genomic instability in the marsh frogs (*Rana ridibunda*) of Sumgayit, Azerbaijan. *Environ. Toxicol. Chem.,* 2005a 24, 2055-2064.

Matson, CW; Palatnikov, G; Islamzadeh, A; McDonald, TJ; Autenrieth, RL; Donnelly, KC; Bickham, JW. Chromosomal damage in two species of aquatic turtles (*Emys orbicularis* and *Mauremys caspica*) inhabiting contaminated sites in Azerbaijan. *Ecotoxicology,* 2005b 14, 513-525.

Matter, B; Schmid, W. Trenimon-induced chromosomal damage in bone-marrow cells of six mammalian species, evaluated by the micronucleus test. Mutat Res, 1971 12, 417-425.

Mersch, J; Beauvias, MN; Nagel, P. Induction of micronuclei in haemocytes and gill cells of zebra mussels, *Dreissena polymorpha*, exposed to clastogens. *Mutat. Res.,* 1996 371, 47-55.

Metcalfe, CD. Induction of micronuclei and nuclear abnormalities in the erythrocytes of mudminnows (*Umbra limi*) and brown bullheads (*Ictalurus nebulosus*). *Bull. Environ. Contam. Toxicol.*, 1988 40, 489-495.

Mouchet, F; Gauthier, L; Mailhes, C; Ferrier, V; Devaux, A. Comparative study of the comet assay and the micronucleus test in amphibian larvae (*Xenopus laevis*) using benzo(*a*)pyrene, ethyl methanesulfonate, and methyl methanesulfonate: establishment of a positive control in the amphibian comet assay. *Environ. Toxicol.,* 2005a 20, 74-84.

Mouchet, F; Gauthier, L; Mailhes, C; Jourdain, MJ; Ferrier, V; Devaux, A. Biomonitoring of the genotoxic potential of draining water from dredged sediments, using the comet and micronucleus tests on amphibian (*Xenopus laevis*) larvae and bacterial assays (Mutatox and Ames tests). *J. Toxicol. Environ. Health* A, 2005b 68, 811-832.

Mouchet, F; Gauthier, L; Mailhes, C; Ferrier, V; Devaux, A. Comparative evaluation of genotoxicity of captan in amphibian larvae (*Xenopus laevis* and *Pleurodeles waltl*) using the comet assay and the micronucleus test. *Environ. Toxicol.,* 2006a 21, 264-277.

Mouchet, F; Baudrimont, M; Gonzalez, P; Cuenot, Y; Bourdineaud, JP; Boudou, A; Gauthier, L. Genotoxic and stress inductive potential of cadmium in *Xenopus laevis* larvae. *Aquat. Toxicol.*, 2006b 78, 157-166.

Mouchet, F; Gauthier, L; Mailhes, C; Jourdain, MJ; Ferrier, V; Triffault, G; Devaux, A. Biomonitoring of the genotoxic potential of aqueous extracts of soils and bottom ash resulting from municipal solid waste incineration, using the comet and micronucleus tests on amphibian (*Xenopus laevis*) larvae and bacterial assays (Mutatox and Ames tests). *Sci. Total Environ.*, 2006c 55, 232-246.

Palhares, D; Grisolia, CK. Comparison between the micronucleus frequencies of kidney and gill erythrocytes in Tilapia fish, following mitomycin C treatment. *Gen. Mol. Biol.*, 2002 25, 281-284.

Pantaleão, SM; Alcântara, AV; Alves, JPH; Spanò, MA. The piscine micronucleus test to assess the impact of pollution on the Japaratuba river in Brazil. *Environ. Mol. Mutagen.*, 2006 47, 219-224.

Rao, SS; Neheli, T; Carey, JH; Cairns, VW. Fish hepatic micronuclei as an indication of exposure to genotoxic environmental contaminants. *Environ. Toxicol. Water Qual.*, 1997 12, 217-222.

Rodriguez-Cea, A; Ayllon, F; Garcia-Vazquez, E. Micronucleus test in freshwater fish species: an evaluation of its sensitivity for application in field surveys. *Ecotoxicol. Environ. Saf.,* 2003 56, 442-448.

Rudek, Z; Rozek, M. Induction of micronuclei in tadpoles of *Rana temporaria* and *Xenopus laevis* by the pyrethroid Fastac 10 EC. *Mutat. Res.,* 1992 298, 25-29.

Saotome, K; Sofuni, T; Hayashi, M. A micronucleus assay in sea urchin embryos. *Mutat. Res.*, 1999 446, 121-127.

Scarpato, R; Migliore, L; Alfinito-Cognetti, G; Barale, R. Induction of micronuclei in gill tissue of *Mytilus galloprovincialis* exposed to polluted marine waters. *Mar. Pollut. Bull.* 1990 21, 74-80.

Schultz, N; Norrgren, L; Grawe, J; Johannisson A; Medhage O. Micronuclei frequency in circulating erythrocytes from rainbow trout (*Oncorhynchus mykiss*) subjected to radiation, an image analysis and flow cytometric study. *Comp. Biochem. Physiol.* C, 1993 105, 207-211.

Shugart, LR. DNA damage as a biomarker of exposure. *Ecotoxicology*, 2000 9, 329-340.

Siboulet, R; Grinfeld, S; Deparis, P; Jaylet, A. Micronuclei in red blood cells of the newt *Pleurodeles waltl* Michah: induction with X-rays and chemicals. *Mutat. Res.*, 1984 125, 275-281.

Siu, WH; Cao, J; Jack, RW; Wu, RS; Richardson, BJ; Xu, L; Lam, PK. Application of the comet and micronucleus assays to the detection of B[a]P genotoxicity in haemocytes of the green-lipped mussel (*Perna viridis*). *Aquat. Toxicol.*, 2004 66, 381-392.

Soldatov, AA. Peculiarities of organization and functioning of the fish red blood system. *J. Evol. Biochem. Physiol.*, 1995 41, 272-281.

Stahl, RG. The genotoxicity of organic compounds in natural waters and waste waters. *Ecotoxicol. Environ. Saf.*, 1991 22, 94-125.

Stegeman, JJ. The cytochromes P450 in fish. In: Hochachka, PW, Mommsen, TP, editors. Molecular Biology Frontiers: Biochemistry and Molecular Biology of Fishes. Amsterdam: Elsevier; 1993; 137-158.

Stich, HF; Tsang, SS; Palcic, B. The effect of retinoids, carotenoids and phenolics on chromosomal instability of bovine papillomavirus DNA-carrying cells. *Mutat. Res.*, 1990 241, 387-393.

Swartz, CD; Donnelly, KC; Islamzadeh, A; Rowe, GT; Rogers, WJ; Palatnikov, GM; Mekhtiev, AA; Kasimov, R; McDonald, TJ; Wickliffe, JK; Presley BJ; Bickham, JW. Chemical contaminants and their effects in fish and wildlife from the industrial zone of Sumgayit, Republic of Azerbaijan. *Ecotoxicology*, 2003 12, 509-521.

Tanzarella, C; Antoccia, A; Ruzza, A. Development of a micronucleus assay that use *Triturus carnifex* to detect genotoxicity of water. In: Bologna, MA, Capula, M, Carpaneto, GM, Luiselli, L, Marangoni, C, Venchi, A, editors. Riassunti del 6° Congresso nazionale della Societas Herpetologica Italica (27 settembre - 1 ottobre 2006). Roma: Stilgrafica, 2007; 64-65.

Tucker, JD; Preston, RJ. Chromosome aberrations, micronuclei, aneuploidy, sister chromatid exchanges, and cancer risk assessment. *Mutat. Res.*, 1996 365, 147-159.

Udroiu, I. Feasibility of conducting the micronucleus test in circulating erythrocytes from different mammalian species: an anatomical perspective. *Environ. Mol. Mutagen.*, 2006a 47, 643-646.

Udroiu, I. The micronucleus test in piscine erythrocytes. *Aquat. Toxicol.*, 2006b 79, 201-204.

Ueda, T; Hayashi, M; Ohtsuka, Y; Nakamura, T; Kobayashi J; Sofuni T. A preliminary study of the micronucleus test by acridin orange fluorescent staining compared with chromosomal aberration test using fish erythropoietic and embryonic cells. *Water Sci. Tech.*, 1992 25, 235-240.

Van Ngan, P; Gomes, V; Passos, MJ; Ussami, KA; Campos, DYF; da Silva Rocha, AJ; Pereira, BA. Biomonitoring of the genotoxic potential (micronucleus and erythrocyte nuclear abnormalities assay) of the Admiralty Bay water surrounding the Brazilian Antarctic Research Station "Comandante Ferraz," King George Island. *Polar. Biol.*, 2007 30, 209-217.

Van Hummelen, P; Zoll, C; Paulussen, J; Kirsch-Volders, M; Jaylet, A. The micronucleus test in *Xenopus*: a new and simple 'in vivo' technique for detection of mutagens in fresh water. *Mutagenesis*, 1989 4, 12-16.

Venier, P; Maron, S; Canova, S. Detection of micronuclei in gill cells and haemocytes of mussels exposed to benzo[a]pyrene. *Mutat. Res.*, 1997 390, 33-44.

Williams, RC; Metcalfe, CD. Development of an in vivo hepatic micronucleus assay with rainbow trout. *Aquat. Toxicol.*, 1992 23, 193-202.

Wilson JT, Dixon DR, Dixon LR. Numerical chromosomal aberrations in the early life-history stages of a marine tubeworm, *Pomatoceros lamarckii* (Polychaeta: Serpulidae). *Aquat. Toxicol.*, 2002 59, 163-175.

Winter, MJ; Ellis, LC; Hutchinson, TH. Formation of micronuclei in erythrocytes of the fathead minnow (*Pimephales promelas*) after acute treatment with mitomycin C or cyclophosphamide. *Mutat. Res.*, 2007 629, 89-99.

Zamora-Perez, A; Zúñiga-González, G; Gómez-Meda, BC; Ramos-Ibarra, ML; Batista-González, C; Torres-Bugarin, O. Induction of micronucleated cells in the shed skin of salamanders (*Ambystoma* sp.) treated with colchicine or cyclophosphamide. *Environ. Mol. Mutagen.*, 2004 44, 436-440.

Zamora-Perez, A; Camacho-Magaña, C; Gómez-Meda, BC; Ramos-Ibarra, ML; Batista-González, C; Zúñiga-González, G. Importance of spontaneous micronucleated erythrocytes in bottlenose dolphin *Tursiops truncatus* to marine toxicology studies *Acta Biol. Hungarica,* 2006 57, 441-448.

Zhuleva, LI; Dubinin, NP. Use of the micronucleus test for assessing the ecological situation in regions of the Astrakhan district. Genetika, 1994 30, 999-1004.

Zuniga-Gonzalez, G; Torres-Bugarin, O; Luna-Aguirre, J; Gonzalez-Rodriguez, A; Zamora-Perez, A; Gomez-Meda, BC; Ventura-Aguilar, AJ; Ramos-Ibarra, ML; Ramos-Mora, A; Ortiz, GG; Gallegos-Arreola, MP. Spontaneous micronuclei in peripheral blood erythrocytes from 54 animal species (mammals, reptiles and birds): part two. *Mutat. Res.,* 2000 467, 99-103.

In: Aquatic Toxicology Research Focus
Editor: Elias P. Svensson, pp. 161-175
ISBN: 978-1-60456-192-0

Chapter 8

COLONISATION OF WATER SYSTEMS IN THE BUILT ENVIRONMENT OF NORTHERN GERMANY BY LEGIONELLA SPP. AND PSEUDOMONAS SPP.

B. P. Zietz*[1] *and H. Dunkelberg
Medical Institute of General Hygiene and Environmental Health,
University of Göttingen, Germany

ABSTRACT

Pneumonia with *Legionella spp.* presents a public health challenge especially because fatal outcomes still remain frequent. *Pseudomonas aeruginosa* is a significant source of hospital-acquired pneumonia and can also cause devastating chronic infections in compromised hosts, for example respiratory infections in cystic fibrosis patients. The aim of our studies was to describe the abundance and epidemiology of *Legionellaceae* in the man-made environment. In total, water systems of 70 different buildings in the German town of Göttingen (Lower Saxony) were examined for the presence of *Legionella* in two sampling cycles. Of these 22 (31%) had the bacterium in at least one water sample. *Legionella pneumophila* serogroups 1, 3, 4, 5 and 6 could be identified in the water samples. Most of the buildings were colonized solely by one distinct strain, as proven by PCR typing. Some buildings contained more than one PCR type or even more than one serogroup. Additionally the colonization of greenhouse misting systems with *Legionella spp.* and *Pseudomonas spp.* was studied in 20 different greenhouse misting systems located in Northern Germany. In total 80 water samples were collected. Each system was tested on two different occasions. Water was drawn at a central tap and at the outlet of spray nozzles. Sampled greenhouses were used to cultivate various plants and trees for commercial, recreational or scientific reasons, some of them in tropical conditions. *Legionella spp.* was detected in 10% of the systems (two systems), but only in low numbers. *Pseudomonas spp.* was recovered from 70% of the greenhouse watering systems (fourteen systems), occasionally at counts greater than 10,000 CFU/100 ml. Each colonized greenhouse had one or several individual strains of *Legionella* and

---

1 Correspondence concerning this article should be addressed to Dr. B. P. Zietz, MPH; Medical Institute of General Hygiene and Environmental Health, University of Göttingen, Lenglerner Str. 75, D-37079 Göttingen, Germany. Tel.: +49 551 5007886-1; fax: +49 551 5007886-3. E-mail: bjoern.zietz@nlga.niedersachsen.de.

Pseudomonas that could not be detected in any other system. This was demonstrated by a random amplified polymorphic DNA typing method. The possible health hazard caused by these water systems for both genera of bacteria is evaluated and discussed.

Keywords: *Legionella*, *Pseudomonas*, drinking water, hot water supply, greenhouse misting systems, molecular typing, PCR, RAPD.

INTRODUCTION

Legionellae are gram-negative bacteria (rods) that require special culture conditions (Fields et al., 2002). They were first recognized as the causative agent of Legionnaire's disease in 1977 following an epidemic of acute pneumonia at an American Legion convention in Philadelphia (Fraser et al., 1977; McDade et al, 1977). The bacterium has been recovered from a wide range of man-made water systems including hot water supplies, cooling towers and whirlpool spas (Breiman and Butler 1998). Surveys of lakes, ponds, streams, and soils have indicated that this bacterium is also a common inhabitant of natural waters (Fliermans, 1996).

Bacteria of the genus *Pseudomonas* are gram-negative rods with polar flagella (Timmis, 2002). Similar to *Legionella* they are common environmental bacteria (Costerton et al., 1999; Stover et. al., 2000). They can be detected in groundwater and drinking water systems (Schoenen et al., 1986; Codony et al., 2002; Banning et al., 2003). *Pseudomonads* and especially *Pseudomonas aeruginosa* are a significant nosocomial pathogens, which for example cause wound infections and bacteraemia in burn victims or urinary tract infections in catheterized patients (Stover et al., 2000). They are also a source of hospital-acquired pneumonia, particularly in patients on respirators (Garau and Gomez, 2003). A chronic respiratory infection with *P. aeruginosa* is typical for patients with cystic fibrosis (Gibson et al., 2003). Additionally outbreaks of *Pseudomonas* dermatitis / folliculitis and otitis externa associated with swimming pool and whirlpool spa use have been described (CDC, 2000; Ratnam et al., 1986).

MATERIALS AND METHODS

Legionellaceae in Warm Water Supplies of Buildings

In a first sampling period, water was collected from 70 private and public buildings between February and September 1999 in the Göttingen area of Germany. In private buildings, water was taken in the bathroom from the hot-water taps. Water samples from public buildings were obtained by turning on the hot-water taps (many from showers heads) and taking the first water. A second sample was taken when the water reached the highest temperature. In the first sampling period, a total of 129 samples were collected (Zietz et al., 2001). In a second sampling period, 196 samples from the same buildings (including additional follow-up samples of colonized buildings) were collected between September 1999 and November 2000 (Brengelmann et al., 2003).

All sampled buildings were supplied by the water plant of Göttingen (Stadtwerke Göttingen). The distribution system consists of three main pressure zones according to different elevations of the city and several small higher or more distant zones. Three facilities that combine water from a transport pipe from the Harz Mountains (about 80%) and their local well supplied water to the lowest zone (Schumacher et al., 1988). Out of this zone water is pumped up to the other zones.

Water samples (1 l volume) were filtered through 0.45-µm-pore-size cellulose-mixed ester filters with a diameter of 50 mm (Schleicher and Schuell, Dassel, Germany) using a vacuum pump. Then 10 ml of a KCl/ HCl-buffer (0.2 M, adjusted to pH 2.2) was poured onto the filter and removed again after 5 minutes. The filters were placed on a MWY agar plate (Oxoid, Wesel, Germany) and incubated at 37°C in a humidified atmosphere (plastic bag) for seven days and examined daily. Additionally 1 ml of water was added to 1 ml of a KCl/ HCl-buffer and after 5 minutes 0.5 ml of the solution was used to inoculate the surface of the MWY agar. This was done in duplicate. Colonies that morphologically matched *Legionella* colonies were subcultured onto blood and MWY agar. Representative colonies (1-2) of those that failed to grow on blood agar were examined by direct fluorescent antibody technique. Isolates were stored at -70°C (Microbank, Mast Diagnostica).

RAPD-Polymerase chain reaction: To identify different strains of *Legionella* we used three different primers to amplify DNA fragments in crude bacterial lysates to generate banding profiles (Wiese et al., 2004; Zietz et al., 2002). Used primers were ERIC2 (5'-AAG TAA GTG ACT GGG GTG AGC G-3', (van Belkum et al., 1993) and a combination of Lpm-1 (5'-GGT GAC TGC GGC TGT TAT GG-3') and Lpm-2 (5'-GGC CAA TAG GTC CGC CAA CG-3') (Jaulhac et al., 1992). ERIC2 is an enterobacterial repetitive intergenic consensus motif. Lpm-1 and Lpm-2 are part of the macrophage infectivity potentiator (mip) gene of *Legionella*. Gels were stained by adding ethidium bromide to the agarose gel and bandings were visualized under ultraviolet light. Details of the methods can be found in Zietz et al. (2001).

LEGIONELLA SPP. AND PSEUDOMONAS SPP. IN GREENHOUSE MISTING SYSTEMS

Water was collected from private and public greenhouse misting systems between June and September 2003 in Lower Saxony and Hessia, Germany. Sampled greenhouses were used to cultivate various plants and trees for commercial, recreational or scientific reasons, some of them in tropical conditions. Some of the greenhouses were open to the public. In total 20 different greenhouse misting systems were tested on two different occasions. Each time, water was drawn at the outlet of spray nozzles and at a central tap of the internal misting system. Therefore in total 80 water samples were collected. Water was not flushed except the first water stagnated in the pipes connecting the tap with the central (circulating) system (amount estimated). Greenhouse misting systems were supplied by public drinking water systems, by rain water cisterns or by private wells. In one case water was supplied from a pond (Zietz et al., 2006).

To detect *Legionella spp*., water samples (100 and 20 ml) were treated as described above with the exception that MWY agar was replaced by GVPC agar. Per sample 5-8 colonies of those that failed to grow on blood agar were examined using a *Legionella spp.* latex test kit.

For the detection of *Pseudomonas spp*., samples of 100 and 20 ml water were vacuum filtered through 0.45-µm-pore-size cellulose-mixed ester filters (Schleicher and Schuell) and placed on Cetrimide agar plates (Merck, Darmstadt, Germany). Additionally, 0.5 ml of water was directly inoculated on the surface of the Cetrimide agar in duplicate. Colonies were examined under UV light for fluorescence and representative suspicious colonies were subcultured onto blood agar. Finally, following a Gram stain diagnosis and a positive oxidase test, the species of the isolated colonies were determined using BBL Crystal Enteric/NF System (Becton Dickinson).

For every sample, numbers of total heterotrophic plate counts per millilitre at 20 and 36°C were determined. The parameters pH, conductivity, total hardness and different ions were determined from an additional water sample at every greenhouse sampling date. For details of the methods please refer to Zietz et al., 2006.

For typing isolates of *Legionella* and *Pseudomonas* the same RAPD-Polymerase chain reaction method was used as described above. Primer 272 (5′-AGC GGG CCA A-3′) (Mahenthiralingam et al., 1996; Ruimy et al., 2001) was used to generate banding profiles for *Pseudomonas spp*. isolates.

RESULTS

Legionellaceae in Warm Water Supplies of Buildings

Of the 70 buildings examined in the first sampling period 18 (26%) had the bacterium in at least one water sample. *Legionella pneumophila* serogroups 1, 4, 5 and 6 could be identified in the water samples (Zietz et al., 2001). In the second period 17 buildings (24%) had positive results in the culture (Brengelmann et al., 2003). Of these, 4 buildings were newly detected to have cultureable *Legionellaceae* compared with the first sampling period. In addition to the serogroups found before, serogroup 3 was detected for the first time (Table A). No other species than *L. pneumophila* was found. In total 31% of the buildings had the bacterium in at least one water sample

The overall trend was that the larger the building's plumbing had been, the more samples were positive for *Legionella* and the more bacteria could be found. The highest detected concentration of bacteria was 78,000 CFU/l in the first period and 150,000 CFU/l in the second period. In the second sampling period 3 buildings were found to have a bacteria concentration above 10^5 CFU/l (Table B).

Table A. Detected *Legionellaceae* in different types of buildings

Type of building	Number of tested buildings	First sampling period		Second sampling period	
		Buildings with at least one sample containing *Legionellaceae*	Found serogroups *	Buildings with at least one sample containing *Legionellaceae*	Found serogroups *
Sports halls and swimming baths	24	5 (21%)	1, 4, 5, 6	4 (17%)	1, 4, 6
University buildings	19	5 (26%)	1, 4, 6	3 (16%)	1, 6
Hospitals and old people's homes	8	5 (63%)	1, 4, 6	7 (88%)	1, 3, 4, 5
Halls of residence	4	2 (50%)	1, 5	2 (50%)	1, 4
Hotels	4	0 (0%)	-	0 (0%)	-
Other buildings	11	1 (9%)	1	1 (9%)	4
Total	70	18 (26%)	1, 4, 5, 6	17 (24%)	1, 3, 4, 5, 6

Footnotes: * all isolates *L. pneumophila*

Table B. The distribution of the maximum found colony forming units (CFU/l) of *Legionella spp.* for all tested buildings

Range of detected colony forming units (CFU/l)	Number of buildings - first sampling period	Number of buildings - second sampling period
Not detectable	52 buildings	53 buildings
$1\text{-}10^2$	7 buildings	5 buildings
$>10^2\text{-}10^3$	1 building	1 building
$>10^3\text{-}10^4$	6 buildings	3 buildings
$>10^4\text{-}10^5$	4 buildings	5 buildings
$> 10^5$	-	3 buildings

Most of the buildings were colonized solely by one distinct strain, as proven by PCR. In three cases equal patterns were found in separate buildings. There were two buildings in this study where isolates with different serogroups were found at the same time.

No association of serogroups or identical PCR types and water supply zones was found (Zietz et al., 2001; Brengelmann et al., 2003).

Legionella Spp. and Pseudomonas SPP. in Greenhouse Misting Systems

Legionella spp. were detected in low numbers in two of the greenhouse misting systems (10%). In the first greenhouse (designated K), *L. pneumophila* serogroup 6 (35 CFU per 100 ml, equivalent to 350 CFU/l) were recovered from the peripheral sample (outlet of spray nozzles) of the first sampling period in July 2003. The central system of the second greenhouse R was found to be colonized by *L. pneumophila* serogroup 6 (6 CFU per 100 ml, equivalent to 60 CFU/l) in the second sampling period. The misting system of greenhouse K was supplied by a private well and the system of greenhouse R by a rainwater cistern that was filled up with public drinking water if there was a lack of rainwater (in the hot summer probably filled up several times during the sampling period, but there was no documentation). In total, 13 isolates have been stored and typed by RAPD. All isolates belonged to serogroup 6 and were recovered from greenhouse K or R. Isolates from each greenhouse had a unique typing pattern and all isolates from the same greenhouse were identical (Zietz et al., 2006).

Pathogenic *Pseudomonas spp.* could be found in 70% of the greenhouse watering systems (14 systems), occasionally in excess of 10^4 CFU per 100 ml. The species *P. aeruginosa*, *P. putida*, *P. fluorescens* and one single isolate of *P. stutzeri* were detected. In three systems, only one of the four samples (2 x central, 2 x peripheral/outlet of spray nozzles) contained *Pseudomonas spp.* In seven systems, *Pseudomonas spp.* were found in two different samples, in two systems in three samples and in one system they were cultured from all four samples.

In total, 59 isolates of *P. aeruginosa* were recovered from the greenhouses, of which 34 different PCR patterns could be identified. Additionally, each of the five *P. aeruginosa* reference strains had an individual pattern. Each greenhouse had one or more individual patterns that could not be detected in any other system. In three cases, a pattern from an isolate of the first sampling cycle could be found again in isolates of the second cycle. *P. fluorescens* was recovered from five different greenhouses, each having one single isolate with its own individual PCR banding pattern. *P. putida* was also detected in five different greenhouses. From one greenhouse, three isolates were recovered and a single isolate from each of the other ones. All greenhouse strains and the reference strain (DSM 291) had their own distinct pattern (Fig. 1). The three isolates of greenhouse M produced an equal banding pattern (isolates PpM1Cb, PpM1P, PpM1Ca) (Zietz et al., 2006).

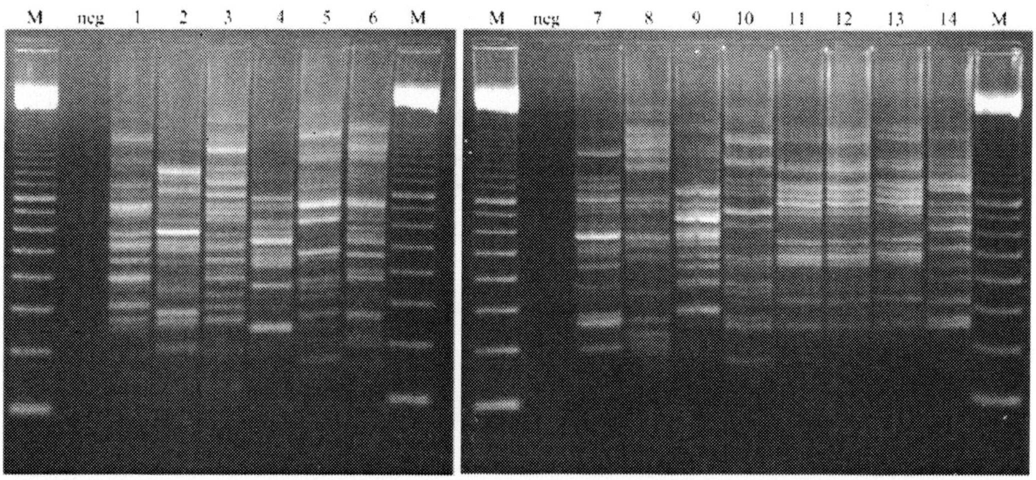

Figure 1. Fingerprint profiles of *P. fluorescens* isolates (left) and *P. putida* isolates (right) from greenhouse misting systems tested with primer sequence 272. Lanes (left to right): molecular weight marker (100 bp), negative control (water blank), *P. fluorescens* reference strain DSM 50090 (lane 1), isolates PfG2C, PfO1C, PfL1P, PfF1C, PfC1C, molecular weight marker. Second part: molecular weight marker (100 bp), negative control, *P. putida* DSM 291 (lane 7), PpL2P, PpC2P, PpN1C, PpM1Cb, PpM1P, PpM1Ca, PpF1C, molecular weight marker.

Total heterotrophic counts at 20 and 36°C varied in a broad range between 0 CFU and $>10^4$ CFU per ml. The average CFUs from peripheral samples were higher than that from central samples. There were frequently significant ($P < 0.05$ or $P < 0.01$) Spearman rank correlations between *Pseudomonas* numbers and total heterotrophic counts (only in second sampling cycle) or between different types of total heterotrophic counts in the same sampling cycle. The results indicate that microbial enumeration of several species can occur simultaneously, probably related to biofilms (Zietz et al., 2006).

DISCUSSION

Legionella spp. in warm water supplies: Also in Lower Saxony, Germany, Habicht and Müller (1988) found that 70% of the 103 hospitals and 18% of the 62 hotels investigated were positive for *Legionella*. These findings are similar to our results testing only one supply area. Boschek et al. (1995) were able to culture *Legionellaceae* from warm water of 11 out of 12 sampled hospitals in a German town. A study in south-eastern Germany investigated *Legionella* contamination in 75 different buildings. In total 85% of all public buildings (hotels, scientific facilities, swimming baths) and 65% of all private households in large buildings had the bacteria in their water samples. Most samples contained *Legionella spp.* in the range of 10^3 - 10^5 CFU/l, the highest concentration was 4 x 10^6 CFU/l. Serogroups 1, 3, 6, and 10 were most commonly isolated with frequencies of 22%, 14%, 16% and 18%, respectively. Of the buildings, 31 were colonized by one serogroup and 15 by two or more. Additionally 5 strains (6%) were non-pneumophila species (Lück et al., 1993). A high constancy (regular detection, only a few minor genetical changes) of *Legionella* colonization was found in a hospital connected with a ring pipe warm water system over 7 years (Lück et al., 1995). It has to be taken into account that used methods varied between these studies.

It is important to know that hospitals and old people's homes used by a risk group for infections frequently had high rates of contamination including the Göttingen study.

No association of serogroups or identical PCR types and water supply zones was found in our study in Göttingen. So the exact evolution and origin of these populations remains unclear. But it seems likely that occasionally *Legionella* colonies from environmental sources are transported through the water supply system to the user's installation. In some cases they are able to colonize especially the hot water systems.

Greenhouse misting systems: Regarding greenhouse contamination only one related study has been published which examined different gardening farms located in eastern Poland (Lublin province) for the presence of *Legionella spp.* in tap water, soil and air (Stojek and Dutkiewicz, 2002). Of the 20 samples taken from indoor taps in traditional greenhouses used for sprinkling on plants, 25.0% were culture positive for *Legionella spp.* In outdoor taps used for sprinkling on plants, 22.2% of the 36 samples contained *Legionella spp.* No bacteria were detected in the tap-water, soil and air samples collected from modern greenhouses.

Legionella spp. infections (especially *L. longbeachae*) have been associated with gardening and the use of potting soil in Australia, Asia, North America and Europe (Cameron et al., 1991; Anonymous, 2000; Suzuki et al., 2002; den Boer et al., 2007). *Legionella spp.* was also detected within slow sand filters used for fungal plant pathogen suppression in horticultural crops (Calvo-Bado et al., 2003) and composted plant materials (Hughes and Steele, 1994).

Our study on greenhouse misting systems indicates that sources other than hot water installations, cooling towers and whirlpool spas should be observed as possibly critical for a relevant *Legionella* contamination. A possible *Pseudomonas* contamination should also be kept in mind especially when untreated water is used in man-made systems.

HEALTH EVALUATION OF LEGIONELLA SPP. CONTAMINATION

The numbers of colony forming units found in greenhouse misting systems (350 and 60 CFU/l, respectively) appear to be relatively low compared with the numbers of *Legionella* detected in our study of hot water systems in public and private buildings in the city of Göttingen. The city is located in centre of the area studied in the greenhouse investigation. In the 70 public and private buildings, up to 150,000 CFU/l of *Legionella spp.* were recovered from hot water taps. Of the 70 buildings examined, 31% had the bacterium in at least one water sample compared to 10% of greenhouse misting systems (Zietz et al. 2001; Brengelmann et al., 2003).

The conditions in these building hot water systems may have been more favourable for *Legionella spp.* growth than in the greenhouse misting systems. The summer of 2003 (June to September) was rather warm in Germany, generally providing good conditions for the enumeration of *Legionella spp.* in greenhouse water systems. Air temperatures often exceeded 25°C during the sampling period. The temperatures of the water samples in the greenhouse misting systems were usually between 20 and 30°C, although there was sometimes a high turnover of water (statement of the system operators) and probably lower soil temperatures. *Legionellaceae* are known to multiply at temperatures between 25 and 42°C, with an optimal growth temperature of 35°C (Fields et al., 2002).

The US Occupational Safety and Health Administration recommends action level 2 ("immediate cleaning and/or biocide treatment; take prompt steps to prevent employee exposure") for *Legionella* concentrations in domestic water exceeding 10^5 CFU/l and action level 1 ("prompt cleaning and/or biocide treatment of the system") for concentrations exceeding 10^4 CFU/l. For *Legionella* concentrations in humidifiers, both of these action levels are only one tenth of the values stated above (OSHA, 1999)

In Germany, the action level for the renovation of existing tap-water systems is regulated by a DVGW technical norm. The most recent norm states that if *Legionella* concentration exceeds 10^3 CFU/l, a renovation and/or cleaning has to be initiated. Concentrations above 10^5 CFU/l are considered extremely high, which make immediate cleaning and prompt steps for exposure prevention necessary (DVGW, 2004). These numbers are technically orientated and are not based on estimates of infective doses. As one can see, the *Legionella* concentrations found in tap-water of several buildings in Göttingen are in the range where prevention steps are necessary. In greenhouse misting systems detected *Legionella* concentrations (350 CFU/l and 60 CFU/l, respectively) are below all of these action levels, but they indicate that these systems are contaminated.

EVALUATION OF POSSIBLE HEALTH EFFECTS OF PSEUDOMONAS SPP. CONTAMINATION

Water of misting systems in our greenhouse study contained *Pseudomonas spp.* in concentrations sometimes higher than 10^4 CFU per 100 ml. High counts were more frequent in central samples and in water collected in the second sampling cycle. In comparison with published data on other water systems, detected numbers of *Pseudomonas spp.* appear high to very high.

For example, the water distribution system of Barcelona and surrounding cities was tested for the presence of members of the *Pseudomonadaceae* family (Ribas et al., 2000). *Pseudomonas spp.* were found in concentrations of up to 74 CFU/100 ml. The authors reported that untreated water contained high numbers of these bacteria. In the Italian city of Bologna shower water of 12 swimming pools contained *P. aeruginosa* in 22 out of 48 samples. Concentrations ranged between 2 and 1500 CFU per 100 ml (Leoni et al., 2001).

Pseudomonas spp. is known to be a common species in the first step of the formation of a biofilm in (model) drinking water distribution systems as well as in later phases of biofilm communities (Martiny et al., 2003; Martiny et al., 2005; Schmeisser et al., 2003). In the biofilm environment they are able to successfully adapt to situations of protozoan grazing (Matz et al., 2004). The genus can be detected throughout the stages of typical water purification systems including filtration, softening, deionization and UV treatment (Penna et al., 2002; Mazzola et al., 2006).

Generally, community-acquired pneumonia (CAP) with *Pseudomonas spp.* in previously healthy adults seems to be relatively rare. Although, *Pseudomonas* CAP should be considered in the differential diagnosis for anyone with a smoking history who presents with rapidly progressive pneumonia (Hatchette et al., 2000). In many cases of *Pseudomonas* pneumonia patients had certain risk factors or comorbidities such as lung diseases. This type of pneumonia trends to have severe courses and outcomes (Lesprit et al., 2003; Ramsdell et al., 2005).

In a case description a community-onset *Pseudomonas* pneumonia was associated with the use of a home humidifier. The 39 year-old-man had a medical history including 25 years of asthma (Harris et al., 1984).

To assess detected *Pseudomonas* concentrations in greenhouse misting systems in relation to results of laboratory animal tests, several factors have to be taken into account. Such factors are for example species differences, susceptibility of exposed persons, differing pathogenicity of strains and differences in ways of exposure. Additionally the amount of aerosol deposited in the lungs and the particle size is important. The particle size is known to be an important determinant of the depth of deposition in the lung (ratio peripheral to central deposition).

In laboratory tests, it is reported that in rats a mortality rate of >50% can be induced by a transtracheal infection with 10^6 CFU of *P. aeruginosa* strain PAO1 (Lesprit et al., 2003). In mice, an intranasal inoculation of either *P. aeruginosa* strain PA103 or strain PA103-29 led to a dose-dependent mortality. While inoculation with 10^4 CFU of strain PA103 did not cause death, mortality rates were between 40% and 80%, following administration of 10^5–10^6 CFU (Schultz et al., 2001). Similar results with the intranasal administration of *P. aeruginosa* strain AC869 were found in mice. The 50% lethal dose with this strain was 2.7×10^7 CFU.

Treatment with 1.61 x 10^7 CFU resulted in slight morbidity and some mortality, generally within 3–4 days after the microbial dose (George et al., 1991). According to a review by Rusin et al. (1997), the oral infectious dose for *P. aeruginosa* in animal and human test subjects ranges between 10^8 and 10^9 CFU.

The volume of deposited aerosol water in the respiratory tract of a person staying in a greenhouse is supposed to be low. The volume of deposited aerosol and the location in the respiratory tract where the water is deposited is dependent on the aerosol size generated by the system.

Comparing different available laboratory mortality and morbidity tests with our highest *Pseudomonas* counts (>10^4 CFU/100 ml), there is probably a safety margin remaining (although the exact maximum CFU numbers in our study are unclear). However a detailed risk assessment is still pending and requires further data and investigation. For immune-compromised patients, this calculation may be different (Kooguchi et al., 1998; Faure et al., 2004).

The German drinking water commission recommends a repetition of sampling in 100 ml if *Pseudomonas spp.* is detected in a drinking water sample. In case of confirmation the contamination should be quantified and the systems should be assessed for the reasons of contamination. Afterwards measures for removal of the contamination should be undertaken (disinfection, flushing) (Umweltbundesamt, 2002).

The WHO has evaluated health hazards of *P. aeruginosa* associated with recreational waters (WHO, 2006). It recommends that for continuously disinfected pools, operational levels of *P. aeruginosa* should be below 1 CFU per 100 ml. In natural spas operating with no residual disinfectant, *Pseudomonas* concentrations should be below 10 CFU/100 ml. "If high counts are found (>100/100 ml), pool operators should check turbidity, disinfectant residuals and pH, resample, backwash thoroughly, wait one turnover and resample. If high levels of *P. aeruginosa* remain, the pool should be closed and a thorough cleaning and disinfection programme initiated. Hot tubs should be shut down, drained, cleaned and refilled." Routine monitoring of *P. aeruginosa* is recommended for public and semipublic hot tubs and natural spas.

CONCLUSION

The presented investigations showed that many water systems of buildings in Northern Germany are contaminated with *Legionellaceae*. This is especially true for larger buildings and more complex hot water systems using recirculation of water. Our studies also indicate that aerosolizing greenhouse watering systems may under certain circumstances be a potential source of *Legionella spp.* or *Pseudomonas spp.* infection in gardeners and visitors.

Typing of cultured isolates of *Legionella spp.* and *Pseudomonas spp.* with RAPD PCR showed that this method can be a useful epidemiological tool to investigate a possible infection as a result of a water system contamination with these bacteria. Another main conclusion of our studies is that there exists a great diversity of *Legionella* and *Pseudomonas* strains in man-made water systems as detectable by culture and PCR typing. In epidemiological studies testing water samples for the presence of *Legionella spp.* and *Pseudomonas spp.* several isolates should be identified from a sample because a co-

colonization of a system with different strains is possible. A non-detection of strains in epidemiological investigations may lead to false conclusions.

Greenhouse misting systems should be part of water management programs that include *Legionella spp.* and *Pseudomonas spp.* monitoring and control. Additionally other previously unconsidered man-made water systems should be systematically assessed in the future to uncover a possible bacterial contamination and infection risk.

ACKNOWLEDGEMENTS

The authors thank F. Brengelmann, J. Ebert, B. Gerhart, S. Luthin and M. Narbe, for their cooperation in the investigations.

REFERENCES

Anonymous. Legionnaires' disease associated with potting soil – California, Oregon, and Washington, May–June 2000. *MMWR Morb Mortal Wkly Rep* 2000, 49, 777–778.

Banning, N; Toze, S; Mee, BJ. Persistence of biofilm-associated *Escherichia coli* and *Pseudomonas aeruginosa* in groundwater and treated effluent in a laboratory model system. *Microbiol.* 2003, 149, 47-55.

Boschek, HJ; Langer, BK; Exner, M. Vorkommen von Legionellen in Warmwasserversorgungssystemen von Krankenhäusern und Pflegeheimen einer deutschen Großstadt. [Abstract]. *Zbl Hyg Umweltmed* 1995, 197, 314-315.

Breiman, RF; Butler, JC. Legionnaires' disease: Clinical, epidemiological, and Public Health perspectives. *Sem. Resp. Inf.* 1998, 13, 84-89.

Brengelmann, F; Zietz, B; Dunkelberg, H. Nachweis von Legionellen in Gebäuden im Bereich einer kommunalen Trinkwasserversorgung. Aachen: Shaker Verlag; 2003.

Calvo-Bado, LA; Morgan, JA; Sergeant, M; Pettitt, TR; Whipps, JM. Molecular characterization of *Legionella* populations present within slow sand filters used for fungal plant pathogen suppression in horticultural crops. *Appl. Environ. Microbiol.* 2003, 69, 533-541.

Cameron, S; Roder, D; Walker, C; Feldheim, J. Epidemiological characteristics of *Legionella* infection in South Australia: implications for disease control. *Aust N Z J Med* 1991, 21, 65-70.

CDC (Centers for Disease Control and Prevention). *Pseudomonas* dermatitis / folliculitis associated with pools and hot tubs--Colorado and Maine, 1999-2000. *MMWR Morb. Mortal Wkly. Rep.* 2000, 49, 1087-1091.

Codony, F; Morato, J; Ribas, F; Mas, J. Effect of chlorine, biodegradable dissolved organic carbon and suspended bacteria on biofilm development in drinking water systems. *J. Basic. Microbiol.* 2002, 42, 311-319.

Costerton, JW; Stewart, PS; Greenberg, EP. Bacterial biofilms: a common cause of persistent infections. *Science* 1999, 284, 1318-1322.

den Boer, JW; Yzerman, EP; Jansen, R; Bruin, JP; Verhoef, LP; Neve, G; van der Zwaluw, K. Legionnaires' disease and gardening. *Clin. Microbiol. Infect.* 2007, 13, 88-91.

DVGW (Deutsche Vereinigung des Gas- und Wasserfaches e.V.). Trinkwassererwärmungs- und Trinkwasserleitungsanlagen; Technische Maßnahmen zur Verminderung des Legionellenwachstums; Planung, Errichtung, Betrieb und Sanierung von Trinkwasser-Installationen. DVGW Arbeitsblatt W 551. Bonn: DVGW; 2004.

Faure, K; Sawa, T; Ajayi, T; Fujimoto, J; Moriyama, K; Shime, N; Wiener-Kronish, JP. TLR4 signaling is essential for survival in acute lung injury induced by virulent *Pseudomonas aeruginosa* secreting type III secretory toxins. *Respir. Res.* 2004, 5, 1.

Fields, BS; Benson, RF; Besser, RE. *Legionella* and Legionnaires' disease: 25 years of investigation. *Clin. Microbiol. Rev.* 2002, 15, 506-526.

Fliermans, CB. Ecology of *Legionella*: From data to knowledge with a little wisdom. *Microb. Ecol.* 1996, 32, 203-228.

Fraser, DW; Tsai, TR; Orenstein, W; Parkin, WE; Beecham, HJ; Sharrar, RG; Harris, J; Mallison, GF; Martin, SM; McDade, JE; Shepard, CC; Brachman, PS. Field Investigation team. Legionnaires' disease: description of an epidemic of pneumonia. *N. Engl. J. Med.* 1977, 297, 1189-97.

Garau, J; Gomez, L. *Pseudomonas aeruginosa* pneumonia. *Curr Opin Infect Dis* 2003, 16, 135-143.

George, SE; Kohan, MJ; Whitehouse, DA; Creason, JP; Kawanishi, CY; Sherwood, RL; Claxton, LD. Distribution, clearance, and mortality of environmental *pseudomonads* in mice upon intranasal exposure. *Appl. Environ. Microbiol.* 1991, 57, 2420-2425.

Gibson, RL; Burns, JL; Ramsey, BW. Pathophysiology and management of pulmonary infections in cystic fibrosis. *Am. J. Respir. Crit. Care Med.* 2003, 168, 918-951.

Habicht, W; Müller, HE. Occurrence and parameters of frequency of *Legionella* in warm water systems of hospitals and hotels in Lower Saxony. *Zbl. Bakt. Hyg B* 1988, 186, 79-88.

Harris, AA; Goodman, L; Levin, S. Community-acquired *Pseudomonas aeruginosa* pneumonia associated with the use of a home humidifier. *Wes.t J. Med.* 1984, 141, 521–523.

Hatchette, TF; Gupta, R; Marrie, TJ. *Pseudomonas aeruginosa* community-acquired pneumonia in previously healthy adults: case report and review of the literature. *Clin Infect Dis* 2000, 31, 1349–1356.

Hughes, MS; Steele, TW. Occurrence and distribution of *Legionella* species in composted plant materials. *App.l Environ. Microbiol.* 1994, 60, 2003-2005.

Jaulhac, B; Nowicki, M; Bornstein, N; Meunier, O; Prevost, G; Piemont, Y; Fleurette, J; Monteil, H. Detection of *Legionella spp.* in bronchoalveolar lavage fluids by DNA amplification. *J. Clin. Microbiol.* 1992, 30, 920-924.

Kooguchi, K; Hashimoto, S; Kobayashi, A; Kitamura, Y; Kudoh, I; Wiener-Kronish, J; Sawa, T. Role of alveolar macrophages in initiation and regulation of inflammation in *Pseudomonas aeruginosa* pneumonia. *Infect. Immun.* 1998, 66, 3164-3169.

Leoni, E; Legnani, PP; Bucci Sabattini, MA; Righi, F. Prevalence of *Legionella spp.* in swimming pool environment. *Water. Res.* 2001, 35, 3749-3753.

Lesprit, P; Faurisson, F; Join-Lambert, O; Roudot-Thoraval, F; Foglino, M; Vissuzaine, C; Carbon, C. Role of the quorum-sensing system in experimental pneumonia due to *Pseudomonas aeruginosa* in rats. *Am. J. Respir. Crit. Care Med* .2003, 167, 1478-1482.

Lück, PC; Leupold, I; Hlawitschka, M; Helbig, JH; Carmienke, I; Jatzwauk, L; Guderitz, T. Prevalence of *Legionella* species, serogroups, and monoclonal subgroups in hot water systems in south-eastern Germany. *Zentralbl. Hyg. Umweltmed.* 1993, 193, 450-460.

Lück, PC; Schütz, M; Helbig, JH; Jatzwauk, L. Konstanz und Variabilität von *Legionella pneumophila*-Populationen im Wassersystem eines Universitätsklinikums. [Abstract]. *Zbl. Hyg. Umweltmed.* 1995, 197, 317-318.

Mahenthiralingam, E; Campbell, ME; Foster, J; Lam, JS; Speert, DP. Random amplified polymorphic DNA typing of *Pseudomonas aeruginosa* isolates recovered from patients with cystic fibrosis. *J. Clin. Microbiol.* 1996, 34, 1129-1135.

Martiny, AC; Albrechtsen, HJ; Arvin, E; Molin, S. Identification of bacteria in biofilm and bulk water samples from a nonchlorinated model drinking water distribution system: detection of a large nitrite-oxidizing population associated with *Nitrospira spp. Appl. Environ. Microbiol.* 2005, 71, 8611-8617.

Martiny, AC; Jorgensen, TM; Albrechtsen, HJ; Arvin, E; Molin, S. Long-term succession of structure and diversity of a biofilm formed in a model drinking water distribution system. *Appl. Environ. Microbiol.* 2003, 69, 6899-6907.

Matz, C; Bergfeld, T; Rice, SA; Kjelleberg, S. Microcolonies, quorum sensing and cytotoxicity determine the survival of *Pseudomonas aeruginosa* biofilms exposed to protozoan grazing. *Environ. Microbiol.* 2004, 6, 218-226.

Mazzola, PG; Martins, AM; Penna, TC. Chemical resistance of the gram-negative bacteria to different sanitizers in a water purification system. *BMC Infect. Dis.* 2006, 6, 131.

McDade, JE; Shepard, CC; Fraser, DW; Tsai, TR; Redus, MA; Dowdle, WR. Legionnaires' disease: isolation of a bacterium and demonstration of its role in other respiratory disease. *N. Engl. J. Med.* 1977, 297, 1197-1203.

OSHA (Occupational Safety and Health Administration). (1999) OSHA Technical Manual (TED 1-0.15A), Section III - Chapter 7. Legionnaires' disease (1999, January 20). Washington, DC: Occupational Safety and Health Administration. [cited 2007, May 22nd]. Available from: URL: http://www.osha.gov/dts/osta/otm/otm_iii/otm_iii_7.html

Penna, VT; Martins, SA; Mazzola, PG. Identification of bacteria in drinking and purified water during the monitoring of a typical water purification system. *BMC Public Health* 2002, 2, 13.

Ramsdell, J; Narsavage, GL; Fink, JB; American College of Chest Physicians, Home Care Network Working Group. Management of community-acquired pneumonia in the home: an American College of Chest Physicians clinical position statement. *Chest* 2005, 127, 1752–1763.

Ratnam, S; Hogan, K; March, SB; Butler, RW. Whirlpool-associated folliculitis caused by *Pseudomonas aeruginosa*: report of an outbreak and review. *J. Clin. Microbiol.* 1986, 23, 655-659.

Rello, J; Bodi, M; Mariscal, D; Navarro, M; Diaz, E; Gallego, M; Valles, J. Microbiological testing and outcome of patients with severe community-acquired pneumonia. *Chest* 2003, 123, 174–180.

Ribas, F; Perramon, J; Terradillos, A; Frias, J; Lucena, F. The *Pseudomonas* group as an indicator of potential regrowth in water distribution systems. *J Appl Microbiol* 2000, 88, 704-710.

Ruimy, R; Genauzeau, E; Barnabe, C; Beaulieu, A; Tibayrenc, M; Andremont, A. Genetic diversity of *Pseudomonas aeruginosa* strains isolated from ventilated patients with nosocomial pneumonia, cancer patients with bacteremia, and environmental water. *Infect. Immun.* 2001, 69, 584–588.

Rusin, PA; Rose, JB; Haas, CN; Gerba, CP. Risk assessment of opportunistic bacterial pathogens in drinking water. *Rev. Environ. Contam. Toxicol.* 1997, 152, 57-83.

Schmeisser, C; Stockigt, C; Raasch, C; Wingender, J; Timmis, KN; Wenderoth, DF; Flemming, HC; Liesegang, H; Schmitz, RA; Jaeger, KE; Streit WR. Metagenome survey of biofilms in drinking-water networks. *Appl. Environ Microbiol.* 2003, 69, 7298-7309.

Schoenen, D; Stoeck, B; Hienzsch, S; Emmel, B. Dekontamination von mit *Pseudomonas aeruginosa* besiedelten Trinkwasserhähnen. *Zentralbl. Bakteriol. Mikrobiol. Hyg. [B]* 1986, 182, 551-557.

Schultz, MJ; Rijneveld, AW; Florquin, S; Speelman, P; Van Deventer, SJ; van der Poll, T. Impairment of host defence by exotoxin A in *Pseudomonas aeruginosa* pneumonia in mice. *J. Med. Microbiol.* 2001, 50, 822-827.

Schumacher, PG; Wagner, I; Kuch, A. Die Trinkwasserversorgung von Göttingen mit Mischwasser. Erfahrungen über den Einfluß der Wasserqualität und von Inhibitoren auf Korrosion im Rohrnetz. *gwf Wasser / Abwasser* 1988, 129, 146-152.

Stojek, NM; Dutkiewicz, J. *Legionella* in sprinkling water as a potential occupational risk factor for gardeners. *Ann. Agric. Environ. Med.* 2002, 9, 261-264.

Stover, CK; Pham, XQ; Erwin, AL; Mizoguchi, SD; Warrener, P; Hickey, MJ; Brinkman, FS; Hufnagle, WO; Kowalik, DJ; Lagrou, M; Garber, RL; Goltry, L; Tolentino, E; Westbrock-Wadman, S; Yuan, Y; Brody, LL; Coulter, SN; Folger, KR; Kas, A; Larbig, K; Lim, R; Smith, K; Spencer, D; Wong, GK; Wu, Z; Paulsen, IT; Reizer, J; Saier, MH; Hancock, RE; Lory, S; Olson, MV. Complete genome sequence of *Pseudomonas aeruginosa* PA01, an opportunistic pathogen. *Nature* 2000, 406, 959-964.

Suzuki, K; Tachibana, A; Hatakeyama, S; Yamaguchi, K; Tateda, K. Clinical characteristics in 8 sporadic cases of community-acquired *Legionella* pneumonia. *Nihon Kokyuki Gakkai Zasshi* 2002, 40, 282-286.

Timmis KN. *Pseudomonas putida*: a cosmopolitan opportunist par excellence. *Environ. Microbiol.* 2002, 4, 779-781.

Umweltbundesamt (Federal Environmental Agency). Empfehlung der Trinkwasserkommission zur Risikoeinschätzung, zum Vorkommen und zu Maßnahmen beim Nachweis von *Pseudomonas aeruginosa* in Trinkwassersystemen. Empfehlung des Umweltbundesamtes nach Anhörung der Trinkwasserkommission des Umweltbundesamtes. *Bundesgesundheitsbl* 2002, 45, 187-188.

van Belkum, A; Struelens, M; Quint, W. Typing of *Legionella pneumophila* strains by polymerase chain reaction-mediated DNA fingerprinting. *J. Clin. Microbiol.* 1993, 31, 2198-2200.

WHO (World Health Organization). Guidelines for safe recreational waters. Volume 2 - Swimming pools and similar recreational-water environments. World Health Organization: Geneva; 2006. [cited 2007, May 22nd]. Available from: URL: http://www.who.int/entity/ water_sanitation_health/bathing/srwe2full.pdf

Wiese, J; Helbig, JH; Lück, PC; Meyer, HG; Jansen, B; Dunkelberg H. Evaluation of different primers for DNA fingerprinting of *Legionella pneumophila* serogroup 1 strains by polymerase chain reaction. *Int. J. Med. Microbiol.* 2004, 294, 401-406.

Zietz, BP; Wiese, J; Brengelmann, F; Dunkelberg, H. Presence of *Legionellaceae* in warm water supplies and identification of strains by polymerase chain reaction. *Epidemiol. Infect.* 2001, 126, 147-152.

Zietz, BP; Dunkelberg, H; Ebert, J; Narbe, M. Isolation and characterization of *Legionella spp.* and *Pseudomonas spp.* from greenhouse misting systems. *J Appl Microbiol* 2006, 100, 1239-1250. Erratum in: *J. Appl. Microbiol.* 2006, 101, 976.

Zietz, BP; Wiese, J; Lück, PC; Helbig, J; Dunkelberg, H. Epidemiological typing of *Legionella pneumophila* serogroup 5 strains. In: *Legionella*. Marre, R; Aabu Kwaik, Y; Bartlett, C; Cianciotto, N; Fields, BS; Frosch, M; Hacker, J; Lück, PC, editors. Washington: ASM Press; 2002; pp. 364-368.

In: Aquatic Toxicology Research Focus
Editor: Elias P. Svensson, pp. 177-188

ISBN: 978-1-60456-192-0

Chapter 9

FERRITIN AND LABILE IRON POOL IN LIMPETS FROM THE BEAGLE CHANNEL

Paula M. González[1], Gabriela Malanga[2] and Susana Puntarulo[1]
[1]Physical Chemistry-PRALIB, School of Pharmacy and Biochemistry, University of Buenos Aires, Buenos Aires, Argentina
[2]Centro Austral de Investigaciones Científicas (CADIC-CONICET), Ushuaia, Tierra del Fuego, Argentina

ABSTRACT

The aim of this work was to study Fe localization and Fe-dependent reactions in digestive glands of two sub-Antarctic limpets *Nacella (P.) magellanica* and *Nacella (P.) deaurata*. In spite of living in the same area, the difference in shore level location affects animal exposure to aerial and environmental conditions. The intertidal species (*N. (P.) magellanica*) is subjected to a variety of stress not experienced by *N. (P.) deaurata*; however *N. deaurata* would be exposed to seawater Fe for longer periods than *N. (P.) magellanica*. Total Fe content in digestive glands was 1.99 ± 0.03 and 0.99 ± 0.07 nmol/mg FW for *N. (P.) deaurata* and *N. (P.) magellanica*, respectively. Lipid radical content was assessed by electronic paramagnetic resonance (EPR), and was higher for *N. (P.) deaurata* than for *N. (P.) magellanica* (192 ± 31 and 122 ± 29 pmol/mg FW, respectively). The labile Fe pool (LIP), assessed by two different techniques, did not showed any difference between both species. Ferritin (Ft) was extracted from digestive glands, purified in a Sephacryl-S-300 column, and the protein was identified by SDS-PAGE. A significantly different Ft content in both species (22.9 and 6.17 ng/mg FW for *N. (P.) deaurata* and *N. (P.) magellanica*, respectively) was observed. The Fe content per molecule of Ft did not show any difference between the species. These results suggest that the Fe excess is efficiently stored in the Ft in *N. (P.) deaurata* since the LIP is not different between species even though *N. (P.) deaurata* showed a higher content of both Fe and Ft, than *N. (P.) magellanica*. However, other mechanisms could be operative in making Fe available for the catalysis of free radical reactions under stress conditions since lipid radical content was higher in *N. (P.) deaurata* as compared to *N. (P.) magellanica*.

INTRODUCTION

Fe has been central to life since the beginning. Environmental dioxygen, a newer component of terrestrial life, was likely to have been the worst environmental pollutant of all times, but evolution has made O_2 so crucial that both Fe and O_2 are absolute requirements for contemporary organisms (Theil, 2003). Fe is an essential element for the growth and well-being of almost all living organisms (Harrison and Arosio, 1996). Fe is involved in many biological functions since by varying the ligands to which it is coordinated, it has access to a wide range of redox potentials and can participate in many electron transfer reactions, spanning the standard redox potential range (Galatro et al., in press). In order to execute its multiple tasks Fe must be incorporated in the heme moieties or alternatively, Fe can bind to enzymes in a form of non-heme moieties or Fe-S motifs (such as those associated with several mitochondrial enzymes) (Cairo et al., 2002). The labile Fe pool (LIP) is defined as a low-molecular-weight pool of weakly chelated Fe that rapidly passes through the cell. It likely consists of both forms of ionic Fe (Fe^{2+} and Fe^{3+}) associated with ligands with low affinity for Fe ions. LIP represents only a minor fraction of the total cellular Fe (3-5%) (Kakhlon and Cabantchik, 2002). It has been proposed that Fe is complexed by diverse low-molecular weight chelators, such as citrate and other organic ions, phosphate, carbohydrates and carboxylates, nucleotides and nucleosides, polipeptides, and phospholipids (Petrat et al., 2002); however, the actual nature of the intracellular ligands participating in LIP formation remains obscure. This Fe is catalytically active in the reactions leading to the generation of extremely reactive species (such as hydroxyl radical) that cause lipid peroxidation, DNA strand breaks and degradation of several biomolecules (Harrison and Arosio, 1996).

Defense against the toxic effect of Fe and O_2 mixtures is provided, in mammals, by two specialized iron-binding proteins: the extracellular transferrins and the intracellular ferritins (Ft). Both proteins retain Fe in the form of Fe^{3+} which, unless mobilized, will not be able to efficiently catalyze the production of free radicals. Fe is stored mainly intracellularly where its potentially damaging effects are greatest. Ft is a nanobox protein designed to contain and maintain in solution up to a few thousands Fe atoms (up to 4500 Fe atoms in mammals, Murgia et al., 2002; Andrews et al., 1992), which otherwise would aggregate in toxic precipitates. Fts originated early in phylogenesis and are present in archeobacteria, eubateria, plants, mammals (Arosio and Levi, 2002), and invertebrates (Charlesworth et al., 1997; Dunkov et al., 1995; von Darl et al., 1994). Ft molecules isolated from vertebrates are composed of two types of subunits (H and L), whereas those from plants and bacteria contain only H-type chains, where H-type is associated with the presence of centers catalyzing the oxidation of two Fe^{2+} atoms (Harrison and Arosio, 1996).

Limpets are very common archaeogastropod molluscs that inhabit intertidal rocky shores. *Nacella (P.) magellanica* and *Nacella (P.) deaurata* are the two most conspicuous limpet species in the Beagle Channel due to their abundance and their relatively large sizes. *N. (P.) magellanica* inhabits the middle and the upper intertidal zones, whereas *N. (P.) deaurata* lives in the lower intertidal and the sublittoral zone (Morriconi, 1999; Morriconi and Calvo, 1993). In spite of living in the same area, the difference in shore level location affects animal exposure to aerial or marine environmental conditions. Due to tidal characteristics of the Beagle Channel, *N. (P.) magellanica* are daily exposed to air twice for 3 to 5 hours each time, but *N. (P.) deaurata* are daily exposed to air for 3 hours, only during spring tides. The different regime of exposition includes extreme temperatures under 0°C during winter time and more than 20°C in summer for *N. (P.) magellanica*, meanwhile *N. (P.) deaurata* are usually covered by more than 0.3 m water at 4°C in winter and 11°C in summer. Thus, *N. (P.) deaurata* would be exposed to Fe seawater for longer periods than *N. (P.) magellanica*. Since the digestive gland is directly affected by the animals feeding status and is also a major target for oxidative disruption related to environmental stress (Malanga et al., 2004; Livingstone, 1991) this study was focussed on the limpets digestive gland to investigate Fe distribution and Fe-dependent oxidative reactions in both sub-Antarctic species.

MATERIALS AND METHODS

1. Collection of Animals

The limpets *Nacella (Patinigera) magellanica* (Gmelin 1971) and *Nacella (Patinigera) deaurata* (Gmelin 1971) were collected from an intertidal area at Punta Occidental (54°50´S, 68°20´W) in the Beagle Channel, in July 2004 (Figure 1). *N. (P.) deaurata* limpets are scarcely exposed to air and were sampled at 0.3-0.5 m water depth in shallow subtidal areas, whereas *N. (P.) magellanica* limpets are daily exposed to air for several hours. Animals had a mean shell length of 51 ± 1 mm and 51 ± 2 mm corresponding to a body fresh weight of 16 ± 1 g and 12 ± 1 g, for *N. (P.) magellanica* and *N. (P.) deaurata*, respectively. No differentiation was made with respect to either sex or reproductive stage. Immediately after collection digestive glands were separated and frozen at -40°C until analyses.

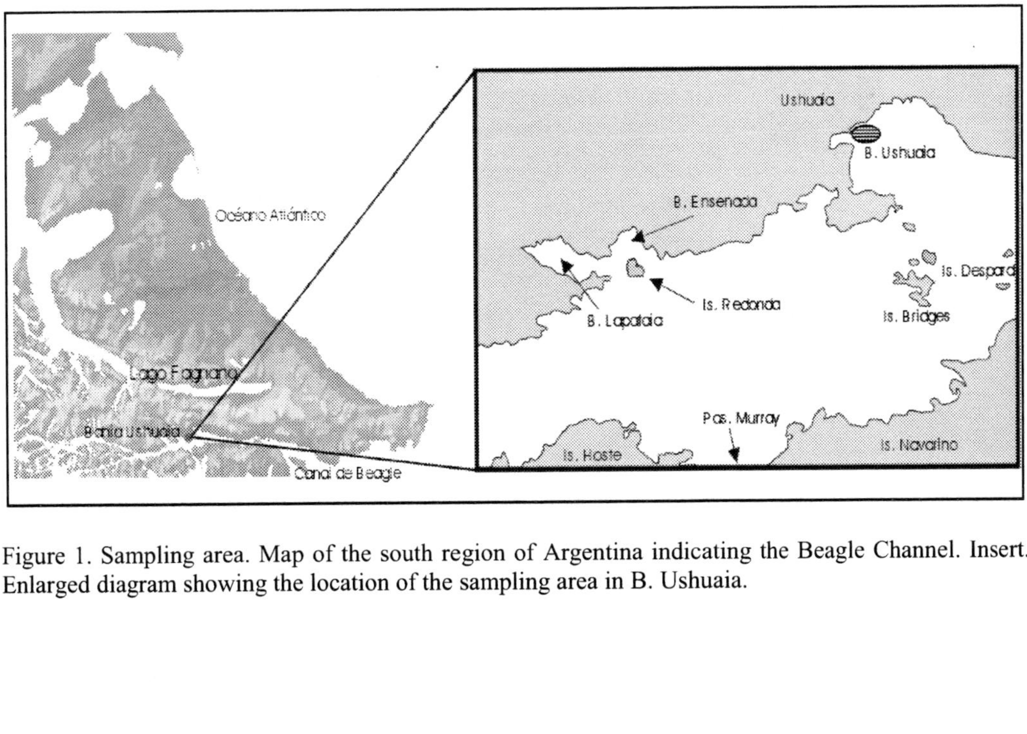

Figure 1. Sampling area. Map of the south region of Argentina indicating the Beagle Channel. Insert. Enlarged diagram showing the location of the sampling area in B. Ushuaia.

2. Fe Content

Isolated digestive glands of the molluscs were digested with a HNO_3 solution. After heating to dryness, the digests were dissolved in 2 ml 5% (v/v) HCl (Lawrie et al., 1991). Concentrations of Fe in the extracts were measured spectrophotometrically after reduction with thioglycolic acid followed by the addition of bathophenanthroline (Brumby and Massey, 1967).

3. Content of Lipid Radical by EPR-Spin Trapping

The homogenates were prepared in potassium phosphate buffer (pH 7.4) containing 50 mM α-(4-pyridyl 1-oxide)-N-t-butyl nitrone (POBN). EPR spectra were obtained at room temperature using a Bruker spectrometer ECS 106, operating at 9.81 GHz with 50 kHz modulation frequency. EPR instrument settings for the spin trapping experiments were: microwave power, 20 mW; modulation amplitude, 1.194 G; time constant, 81.92 ms; receiver gain, 2×10^4 (Jurkiewicz and Buettner, 1994). Quantification was performed according to Kotake et al. (1996).

4. Ft Isolation

Ft from the digestive glands was isolated according to van Gelder et al. (1996) with modifications. The samples were homogenized in a blender with 7 volumes of extraction buffer (10 mM sodium phosphate buffer, 0.15 M NaCl, 1% (w/w) phenylmethanesulfonyl fluoride, pH 7.4) at 4°C. This homogenate was centrifuged at 3000g for 45 min. The supernatent was ultracentrifuged at 78000g for 1 h. The pellet was resuspended in the extraction buffer and centrifuged at 4000g at 4°C for 1 h. These two centrifugation processes were repeated twice and the resuspended pellets were loaded on a Sephacryl S-300 (30 x 1.6 cm) column previously equilibrated in 10 mM potassium phosphate buffer, 0.15 M NaCl, pH 7.0. Fractions were collected and protein and Fe concentrations were assessed. The Fe-rich fractions were pooled and centrifuged at 4°C for 1 h at 78000g. The Ft pellets were disolved in 10 mM potassium phosphate buffer, 0.15 M NaCl, pH 7.0 and stored at 4°C until use. Protein content in the sample was measured according to Bradford (1976) using bovine albumin (Sigma) as standard. Ft Fe content was assessed according to Brumby and Massey (1967).

5. Ft Identification

Ft extraction purity from digestive glands of the limpets was verified by electrophoresis in an SDS-page (Stacking gel: 4% T, 75 V, 20 min. Resolution gel: 14% T, 150 V, 50 min; on denaturating conditions). Gels were stained with coomassie blue to reveal the Ft bands. Comercial Ft from horse spleen extraction was used to identify the Ft bands.

6. LIP Measurement

LIP was determined by both, an EPR technique at 77K according to Woodmansee and Imlay (2002), and by a fluorescence technique with the Fe sensor calcein (CA) according to Darbari et al. (2003) with modifications. For the EPR measurements 400 mg of digestive glands were homogenized in 20 mM Tris-HCl buffer, 135 mM KCl pH 7.4, in Fe free water (IFW) and 2 mM deferoxamine mesylate. The measurements were performed using the following instrument settings: modulation frequency 50 kHz, microwave power 20 mW, microwave frequency 9.44 GHz, centered field 1600 G, time constant 81.92 ms, receiver gain 1×10^4, modulation amplitude 4.753 G and sweep width 800 G. For the fluorescence technique, 250 mg of digestive glands were homogenized in 2 ml of 40 mM potassium phosphate buffer, 120 mM KCl, pH 7.4, in IFW. The homogenate was centrifuged at 10000g for 15 min at 4°C. The supernatant was passed through filters with 30000 Da nominal molecular weight limit. The filtered solution was then reduced with 1 volume of 0.16% thioglycolic acid. Fe in the reduced solution was measured using 1 μM CA solution in 40 mM potassium phosphate buffer, 120 mM KCl, pH 7.4, in IFW. When Fe is added to CA solution a fraction of the dye binds free Fe^{2+} leading to the generation of the Fe-bound (quenched) form [CA-Fe], while another fraction remains free as unbound calcein [CA] and provides the basal fluorescence. The fluorescence (λ_{exc} = 485 nm, λ_{em} = 535 nm) was monitored until stabilization of the signal (F_1) and then deferoxamine mesylate salt was added to a final concentration of 1 mM. The fluorescence was monitored until new stabilization of the signal (F_2). The magnitude of the absolute change in fluorescence (F_2-F_1) is equivalent to the amount of Fe bound to CA. The fractional increase of fluorescence (ΔF), was calculated according to equation 1,

$$\Delta F = (F_2 - F_1) / F_2 \qquad (1)$$

LIP was assessed according to Robello et al. (2007) using equation 2, assuming a dissociation constant (Kd) value of 0.46 μM.

$$LIP = (\Delta F \times [CA]) + [(Kd \times \Delta F) / (1-\Delta F)] \qquad (2)$$

7. Statistical Analyses

Data in the text, figures and tables are expressed as mean ± S.E. of six independent experiments, with two replicates in each experiment. Statistical tests were carried out using Statview for Windows, ANOVA, SAS Institute Inc., version 5.0.

RESULTS

Fe^{2+} ions are able to amplify the damaging effect of reactive O_2 species on membranes, proteins and nucleic acids (Puntarulo and Cederbaum, 1988). To prevent this risk, the cells have evolved two strategies, namely scavenging of hydrogen peroxide or sequestration of Fe

(Zancani et al., 2004). To study the possible role of Fe in the catalysis of lipid radical generation, the Fe content in the digestive gland of both molluscs was examined. Total Fe content was 47.9% lower in *N. (P.) magellanica* as compared to *N. (P.) deaurata* (Figure 2). Besides playing a key role in major intracellular processes and being a strict requirement for life, Fe is an effective catalyst for lipid peroxidation (Puntarulo and Cederbaum, 1988). The initiation reaction of lipid peroxidation is indicated by reaction 3, where one proton is abstracted.

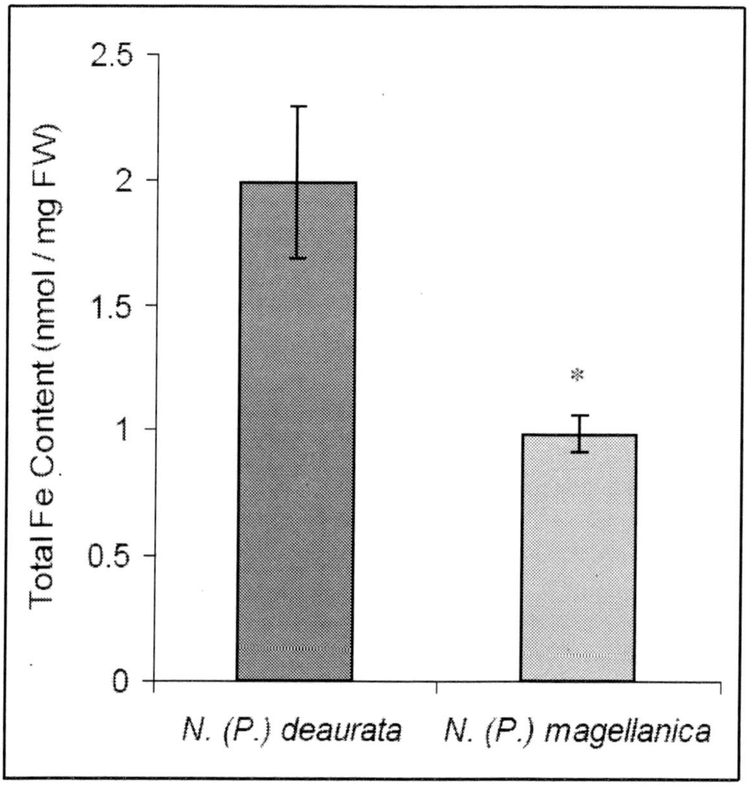

Figure 2. Total Fe content in digestive glands of *N. (P.) deaurata* and *N. (P.) magellanica*. Data are expressed as means ± S.E.M. of six independent experiments. *significantly different at $p \leq 0.05$ from the value obtained for *N. (P.) deaurata*.

$$\text{Lipid} \xrightarrow[\text{- }H^+]{Fe^{2+}} \text{Lipid radical } (LR^\bullet) \qquad (3)$$

The role of Fe and superoxide anion in the initiation step of lipid peroxidation have been extensively discussed (Ursini et al., 1989; Puntarulo and Cederbaum, 1988; Aust et al., 1985). Lipid peroxidation, assessed as the content of $LR^\bullet$ measeured by EPR (Figure 3A), is lower in *N. (P.) magellanica* as compared to *N. (P.) deaurata* (Figure 3B). $LR^\bullet$ content correlates with lower content of transition metals, in agreement with previous observations (Malanga et al., 2004; Estévez et al., 2002).

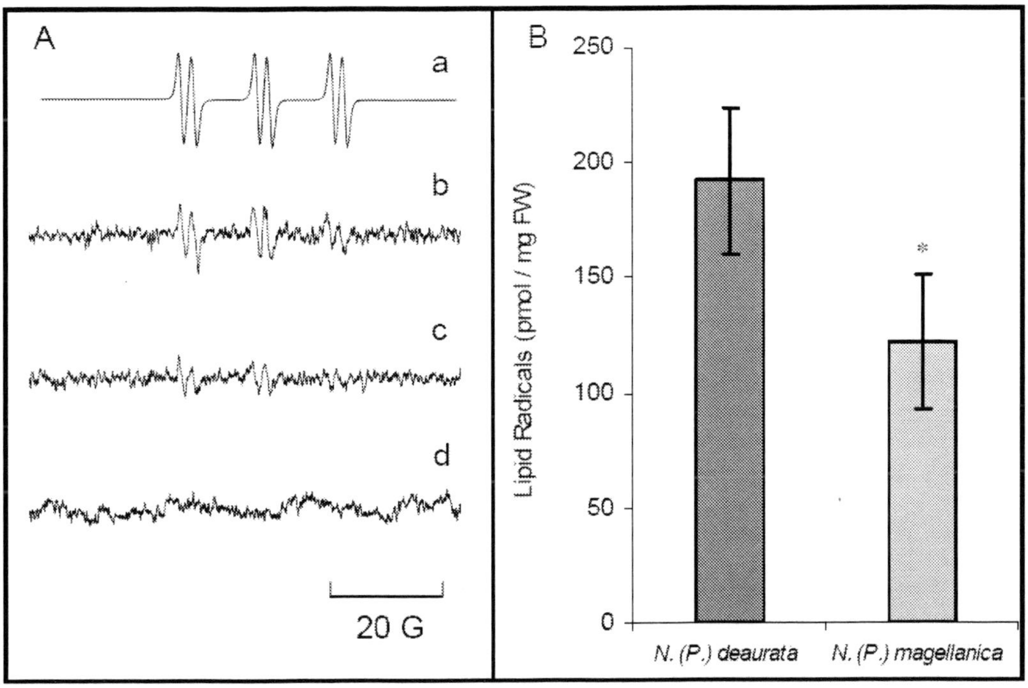

Figure 3. LR$^{\bullet}$ content in digestive glands from the molluscs: A. EPR detection of: (a) computer-simulated EPR spectra exhibiting hyperfine splittings that are characteristic of POBN/LR$^{\bullet}$, (b) typical EPR spectra from *N. (P.) deaurata* digestive gland, (c) typical EPR spectra from *N. (P.) magellanica* digestive gland, and (d) EPR spectra of α-(4-pyridyl 1-oxide)-N-t-butyl nitrone (POBN) itself, a_N = 15.56 G and a_H = 2.79 G (d), are shown. B. LR$^{\bullet}$ content in the digestive glands of *N. (P.) deaurata* and *N. (P.) magellanica* obtained from the EPR measurements. Data are expressed as means ± S.E.M. of six independent experiments. *significantly different at $p \leq 0.05$ from the value obtained for *N. (P.) deaurata*.

On the other hand, the lower LR$^{\bullet}$ steady state concentration in *N. (P.) magellanica* as compared to *N. (P.) deaurata* could be ascribed to either endogenous or exogenous factors. Among the endogenous factors, either a lower concentration of LIP or a lower intracellular conversion rate of bound to bioavailable forms of Fe in *N. (P.) magellanica* as compared to *N. (P.) deaurata*, could lead to a LR$^{\bullet}$ lower content. Even though the actual nature of the intracellular ligands participating in LIP formation remains obscure, it is accepted that the affinity of a chelator for Fe, its ability to permeate different cell compartments as well as the steric accessibility of specific metal binding sites on proteins could all influence the LIP (Galatro et al., in press). A continuous demand for Fe to be available for synthesis of Fe-containing proteins forces a permanent Fe flux from the extracellular milieu (environment) to the cytoplasm and the nature and magnitude of the Fe trafficking in the tissues is a main point to be clarified since it could be the clue to the understanding of the metabolic pathways leading to oxidative stress. EPR detection of LIP by the signal of the deferoxamine-Fe^{3+} complex is understood as a reliable method for detection, but it is not highly sensitive, in opposition to the fluorescence method employing CA, that is a very sensitive approach

(Tarpey et al., 2004). LIP was measured by both methods to confirm the accuracy of the data to be compared. The typical EPR signal of deferoxamine-Fe^{3+} complex was obtained with digestive glands of both limpets (Figure 4A). Quantification of the LIP by the EPR measurements (0.38 ± 0.05 and 0.40 ± 0.05 nmol/mg FW for *N. (P.) deaurata* and *N. (P.) magellanica*, respectively) did not show any significant difference between both species. In the same direction, data in Figure 4B indicate that LIP assessed by CA assay, did not show any significant difference between both species. Thus, both performed analyses confirmed that the total amount of Fe available for catalysis did not differ in both limpets.

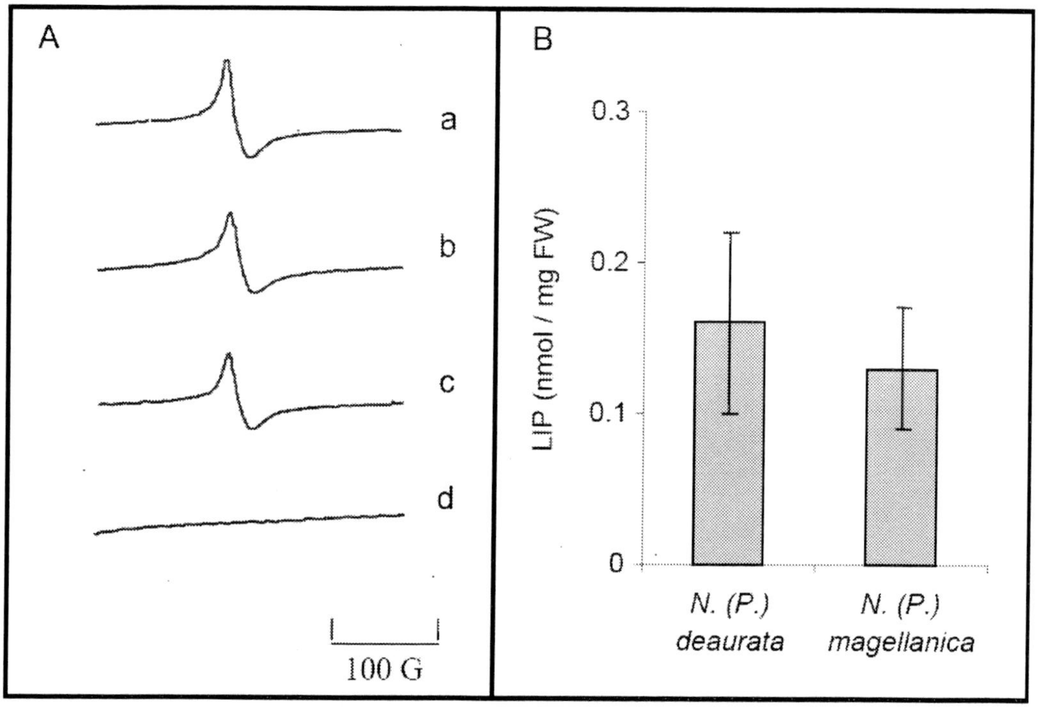

Figure 4. LIP in digestive glands from the molluscs. A. LIP determination by EPR detection of: (a) 355 μM deferoxamine mesylate-Fe^{3+}, (b) *N. (P.) deaurata* digestive gland, (c) *N. (P.) magellanica* digestive gland, and (d) deferoxamine mesylate alone. B. LIP quantification by a fluorescence technique with the Fe sensor CA in the digestive glands of *N. (P.) deaurata* and *N. (P.) magellanica.* Data are expressed as means ± S.E.M. of six independent experiments.

Besides the LIP critical importance as initiator of free radical reactions and the decisive requirement of keeping Fe concentration as low as possible to minimize cellular deterioration, the role of other soluble and insoluble Fe-storage proteins, the formation and contribution of Fe-nitrosyl complexes, glutathione, nitric oxide, etc. should be considered as possible candidates to handle Fe transport and storage under stress conditions. Since Ft has been described as the main storage protein for Fe, experimental isolation of Ft from the digestive glands of both limpets was performed. Digestive glands from *N. (P.) magellanica* yielded a significantly lower amount of Ft (27%), expressed as ng Ft/mg FW, as compared to *N. (P.) deaurata*, and no significantly different Fe content by molecule of Ft between them was observed (Table 1). Subunits analyses of the protein by SDS-polyacrilamide electrophoresis gel indicated that Ft was composed by 21 kDa protein subunits (Figure 5); however, higher molecular mass subunits (30-31 kDa) could appear. The data obtained with horse spleen Ft (Figure 5) showed subunits of 20-21 kDa. The limpets higher subunits could be understood either as a consequence of association processes occurring during the isolation procedure, as products of the association of peptides generated *in vivo* after Ft degradation by proteinases acting on Ft molecules damaged by radicals during Fe exchange, or as a presence of other Ft sequences (Harrison and Arosio, 1996). Further analyses are required to assess this issue.

Table 1. Ft and Fe content by molecule of Ft isolated from digestive glands of *N. (P.) deaurata* and *N. (P.) magellanica*

	N. (P.) deaurata	*N. (P.) magellanica*
Ferritin Content (ng Ft/mg FW)	22.9 ± 5.7	6.2 ± 2.0 *
Ferritin Fe Content (Fe atoms/Ft molecule)	2277 ± 1137	3191 ± 2198

Data are expressed as means ± S.E.M. of two independent experiments.
*significantly different at $p \leq 0.05$ from the value obtained for *N.(P.) deaurata*.

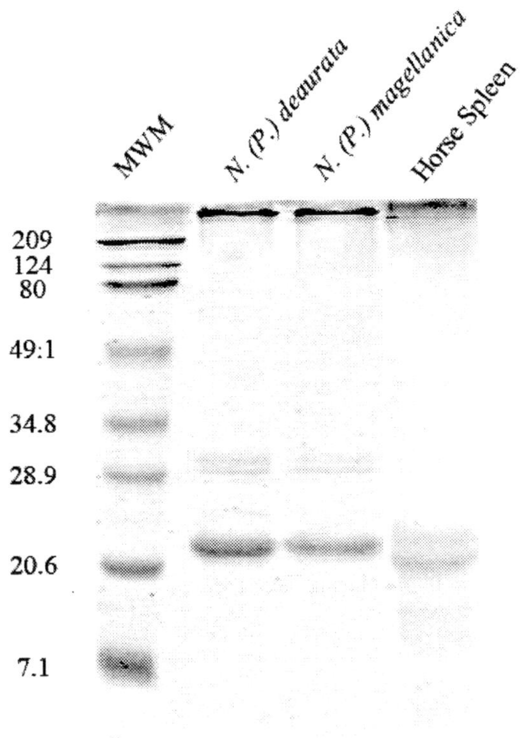

Figure 5. SDS gel at 14 % showing *N. (P.) deaurata*, *N. (P.) magellanica* and horse spleen Ft (comercial) bands.

Fe^{3+} is reduced to Fe^{2+} to be incorporated into Ft where it is stored as Fe^{3+} (*inert* form). By redox reaction with other cellular components, Fe^{3+} from Ft can be re-reduced to Fe^{2+}, released to the cytoplasm and become a catalyst in Fenton-type reactions. Previous data from our laboratory (Malanga et al., 2004) indicated that the maxima Fe^{3+} reduction rate was not significantly different in homogenates of digestive glands of the two molluscs. Thus, this pathway does not seem to be the cause of the differential content of $LR^{\bullet}$ between digestive glands from both limpets. Under stress conditions the steady state concentration of reactive species, such as superoxide anion, could be significantly increased as compared to non-stress conditions (Abele et al., 2007; Abele and Puntarulo, 2004; Heise et al., 2003). This increase in superoxide radical steady state concentration could lead to an increase on Fe release rate from Ft that could be responsible for triggering lipid peroxidation under those stress conditions. This hypothesis would explain the key role of the environmental conditions in the differential developing of oxidative damage in these sub-Antarctic species.

CONCLUSION

Taken as a whole, these data strongly suggest that Fe storage in Ft is a crucial mechanism for preventing the occurrence of radical-driven reactions catalyzed by Fe. However, by the effect of both, exogenous factors such as the exposure to Fe seawater for longer periods, and endogenous factors, such as the significant higher amount of Fe stored in Ft would favored an increased of $LR^{\bullet}$ generation in the membranes of *N. (P.) deaurata* as compared to *N. (P.) magellanica* after exposure to natural stress conditions. The development of stress at the cellular level, triggered by the environment, could be responsible for the release of Fe from Ft through the action of superoxide anion involved in redox reactions. This released Fe could go to the LIP and be available for catalyzing free radical reactions, including lipid peroxidation. Thus, the higher amount of Fe stored in *N. (P.) deaurata* Ft as compared to *N. (P.) magellanica* Ft, may be the key factor for its higher $LR^{\bullet}$ content, even though LIP content is similar in both species under non-stress conditions. Future studies will be required to explore the possible role of other factors, i.e. lipid content and lipid saturation, as a cause of the differential lipid peroxidation in both limpets besides the availability of Fe in Ft.

ACKNOWLEDGMENTS

This study was supported by grants from the University of Buenos Aires (B017), ANPCyT (PICT 11187) and CONICET (PIP 5706). S.P. and G.M. are career investigator from CONICET and P.M.G. is a fellow from the University of Buenos Aires. The authors are grateful to Mr. Daniel Aureliano, Mr. Marcelo Gutierrez and Mrs. Marcela Riggio for their help with the sample collection:

REFERENCES

Abele, D; Philipp, E; González, PM; Puntarulo, S. Marine invertebrate mitochondria and oxidative stress. *Front. Biosci.* (2007) 12: 933-946.

Abele, D.; Puntarulo, S. Formation of reactive species and induction of antioxidant defence systems in polar and temperate marine invertebrates and fish. *Comp. Biochem. Physiol.* (2004) 138A: 405-415.

Andrews, SC; Harrison, PM; Yewdall, SJ; Arosio, P; Levi, S; Bottke, W; von Darl, M; Briat, JF; Laulhère, JP; Lobréaux, S. Structure, function, and evolution of ferritins. *J. Inorg. Biochem.* (1992) 47: 161-174.

Arosio, P; Levi, S. Ferritin, iron homeostasis, and oxidative damage. Free Radic. Biol. Med. (2002) 33: 457-463.

Aust, SD; Morehouse, LA; Thomas, CE. Role of metals in oxygen radical reactions. *J. Free Radic. Med.* (1985) 1: 3-25.

Bradford, MM. A rapid and sensitive method for the quantitation of microgram quantities of protein utilizing the principle of protein–dye binding. *Anal. Biochem.* (1976) 72: 248-254.

Brumby, PE; Massey, V. Determination of nonheme iron, total iron, and copper. *Meth. Enzymol.* (1967) 10: 463-472.

Cairo, G; Recalcati, S; Pietrangelo, A; Minotti, G. The iron regulatory proteins: targets and modulators of free radical reactions and oxidative damage. *Free Radic. Biol. Med.* (2002) 32: 1237-1243.

Charlesworth, A; Giorgieva, T; Gospodov, I; Law, J; Dunkov, B; Ralcheva N; Barillas-Mury, C; Ralchev, K; Kafatos, F. Isolation and properties of *Drosophila melanogaster* ferritin. *Eur. J. Biochem.* (1997) 247: 470-485.

Crichton, RR; Ward, RJ. Iron metabolism – new perspectives in view. *Biochemistry* (1992) 31: 11255-11264.

Darbari, D; Loyevsky, M; Gordeuk, V; Kark, JA; Castro, O; Rana, S; Apprey, V; Kurantsin-Mills, J. Fluorescence measurements of the labile iron pool of sickle erythrocytes. *Blood* (2003) 102: 357-364.

von Darl, M; Harrison, P; Bottke, W. cDNA cloning and deduced amino acid secuence of two ferritin: soma ferritin and yolk ferritin from the snail *Lymnaea stagnalis* L.. *Eur. J. Biochem.* (1994) 222: 353-361.

Dunkov, B; Zhang, D; Choumarov, K; Winzerling, J; Law, J. Isolation and characterization of mosquito ferritin and cloning of a cDNA that encodes one subunit. *Arch. Insect Biochem. Physiol.* (1995) 29: 293-302.

Estévez, MS; Abele, D; Puntarulo, S. Lipid radical generation in polar (*Laternula elliptica*) and temperate (*Mya arenaria*) bivalves. *Comp. Biochem. Physiol.* (2002) 132B: 729-737.

Galatro, A; Rousseau, I; Puntarulo, S. Ferritin role in iron toxicity in animals and plants. *Curr. Top. Toxicol.* (2007). In press.

van Gelder, W; Huijskes-Heis, MIE; Klepper, D; Van Noort, WL; Cleton-Soeteman, MI; Van Eijk, HG. Isolation and partian characterization of two porcine spleen ferritin fractions with different electrophoretic mobility. *Comp. Biochem. Physiol.* (1996) 115B: 191-196.

Harrison, PM; Arosio, P. The ferritins: molecular properties, iron storage function and cellular regulation. *Biochim. Biophys. Acta* (1996) 1275: 161-203.

Heise, K; Puntarulo, S; Portner, HO; Abele, D. Production of reactive oxygen species by isolated mitochondria of the Antarctic bivalve *Laternula elliptica* (King and Broderip) under heat stress. *Comp. Biochem. Physiol.* (2003) 134C: 79-90.

Jurkiewicz, BA.; Buettner, GR. Ultraviolet light-induced free radical formation in skin: an electron paramagnetic resonance study. *Photochem. Photobiol.* (1994) 59: 1-4.

Kakhlon, O; Cabantchik, ZI. The labile iron pool: characterization, measurement, and participation in cellular processes. *Free Radic. Biol. Med.* (2002) 33: 1037-1046.

Kotake, Y; Tanigawa, T; Tanigawa, M; Ueno, I; Allen, DR; Lai, C. Continuous monitoring of cellular nitric oxide generation by spin trapping with an iron–dithiocarbamate complex. *Biochim. Biophys. Acta* (1996) 1289: 362-368.

Lawrie, S; Tancock, N; Mc Growth, S; Roger, J. Influence of complexation on the uptake by plants of iron, manganese, copper and zinc. I. Effect of EDTA in a multimetal and computer simulation study. *J. Exp. Bot.* (1991) 42: 509-515.

Livingstone, DR. Organic xenobiotic metabolism in marine invertebrates. *Adv. Comp. Environ. Physiol.* (1991) 7: 45-185.

Lobréaux, S; Briat, JF. Ferritin accumulation and degradation in different organs of pea (*Pisum sativum*) during development. *Biochem. J.* (1991) 274: 601-606.

Malanga, G; Estévez, MS; Calvo, J; Puntarulo, S. Oxidative stress in limpets exposed to different environmental conditions in the Beagle Channel. *Aquat. Toxicol.* (2004) 69: 299-309.

Morriconi, E. Reproductive biology of limpet *Nacella* (*P.*) *deaurata* (Gmelin 1971) in Bahía Lapataia (Beagle Channel). *Sci. Mar.* (1999) 63: 417-426.

Morriconi, E; Calvo, J. Influencia ambiental en el crecimiento alométrico de la valva de *Nacella* (*P.*) *deaurata* (Gmelin 1971) en el Canal Beagle. *Malacología* (1993) 35: 1-6.

Murgia, I; Delledonne, M; Soave, C. Nitric oxide mediates iron-induced ferritin accumulation in *Arabidopsis*. *Plant J.* (2002) 30: 521-528.

Petrat, F; De Groot, H; Sustmann, R; Rauen, U. The chelatable iron pool in living cells: a methodically defined quantity. *Biol. Chem.* (2002) 383: 489-502.

Puntarulo, S; Cederbaum, AI. Increased NADPH-dependent chemiluminescence by microsomes after chronic ethanol consumption. *Arch. Biochem. Biophys.* (1988) 266: 435- 445.

Robello, E; Galatro, A; Puntarulo, S. Iron role in oxidative metabolism of soybean axes upon growth Effect of iron overload. *Plant Sci.* (2007) 172: 939-947.

Sczekan, SR; Joshi, JG. Isolation and characterization of ferritin from Soyabeans (*Glycine max*). *J. Biol. Chem.* (1987) 262: 13780-13788.

Tarpey, MM; Wink, DA; Grisham, MB. Methods for detection of reactive metabolites of oxygen and nitrogen: *in vitro* and *in vivo* considerations. *Am. J. Physiol. Regul. Integr. Comp. Physiol.* (2004) 286: R431-R444.

Theil, EC. Ferritin: at the crossroads of iron and oxygen metabolism. *J. Nutr.* (2003) 133: 1549S-1553S.

Ursini, F; Maiorano, M; Hochtein, P; Ernster, L. Microsomal lipid peroxidation: mechanisms of initiation. The role of iron and iron chelators. Free Radic. Biol. Med. (1989) 6: 31-36.

Woodmansee, AN; Imlay, JA. Quantitation of intracellular free iron by electron paramagnetic resonance spectroscopy. *Meth. Enzymol.* (2002) 349: 3-9.

Zancani, M; Peresson, C; Biroccio, A; Federici, G; Urbani, A; Murgia, I; Soave, C; Micali, F; Vianello, A; Macri, F. Evidence for the presence of ferritin in plant mitochondria. *Eur. J. Biochem.* (2004) 271: 3657-3664.

INDEX

A

C

D

E

F

G

H

M

N

O

P

Q

R

S